Black Holes
and Baby Universes
and Other Essays

BLACK HOLES
and
BABY UNIVERSES
and Other Essays

Stephen Hawking

BANTAM BOOKS
NEW YORK · TORONTO · LONDON · SYDNEY · AUCKLAND

Black Holes and Baby Universes
and Other Essays

A Bantam Book / October 1993

"Is the End in Sight for Theoretical Physics?", an Inaugural Lecture given in the University of Cambridge by Stephen Hawking. Copyright © 1980, Cambridge University Press. Reprinted by permission.

The interview Desert Island Discs is published through the courtesy of the BBC and with the approval of Mrs. Diana Plomley and Miss Sue Lawley.

Book design by Richard Oriolo

Library of Congress Cataloging-in-Publication Data

Hawking, S. W. (Stephen W.)
  Black holes and baby universes and other essays / Stephen Hawking.
    p.   cm.
  Includes index.
  ISBN 0-553-09523-4 : $21.95
  1. Hawking, S. W. (Stephen W.)   2. Cosmology.   3. Science—
Philosophy.   4. Physicists—Great Britain—Biography.   I. Title.
QC16.H33A3 1993
530.1—dc20   93-8269
                                                                CIP

Published simultaneously in the United States and Canada

Bantam Books are published by Bantam Books, a division of Bantam Doubleday Dell Publishing Group, Inc. Its trademark, consisting of the words "Bantam Books" and the portrayal of a rooster, is Registered in U.S. Patent and Trademark Office and in other countries. Marca Registrada. Bantam Books, 1540 Broadway, New York, New York 10036.

PRINTED IN THE UNITED STATES OF AMERICA

FFG   0   9   8   7   6   5   4

Contents

Contents

PREFACE

T HIS VOLUME CONTAINS a collection of pieces that I wrote over the period 1976 to 1992. They range from autobiographical sketches through the philosophy of science to attempts to explain the excitement I feel about science and the universe. The volume concludes with the transcript of a *Desert Island Discs* program on which I appeared. This is a peculiarly British institution in which the guest is asked to imagine himself or herself cast away on a desert island and is invited to choose eight records with which to while away the time until rescued. Fortunately, I didn't have too long to wait before returning to civilization.

Because these pieces were written over a period of sixteen years, they reflect the state of my knowledge at the time, which I hope has increased over the years. I have therefore given the date and occasion for which each was composed. As each was meant to be self-contained, there is inevitably a certain amount of repetition. I have tried to reduce it, but some remains.

A number of the pieces in this volume were designed to be spoken. My voice used to be so slurred that I had to give lectures and seminars through another person, usually one of my research students who could understand me or who read a text I had written. However, in 1985 I had an operation that removed my powers of speech altogether. For a time I was without any means of communication. Eventually I was equipped with a computer system and a remarkably good speech synthesizer. To my surprise, I found I could be a successful public speaker, addressing large audiences. I enjoy explaining science and answering questions. I'm sure I have a lot to learn about how to do it better, but I hope I'm improving. You can judge for yourselves whether I am by reading these pages.

I do not agree with the view that the universe is a mystery, something that one can have intuition about but never fully analyze or comprehend. I feel that this view does not do justice to the scientific revolution that was started almost four hundred years ago by Galileo and carried on by Newton. They showed that at least some areas of the universe do not behave in an arbitrary manner but are governed by precise mathematical laws. Over the years since then, we have extended the work of Galileo and Newton to almost every area of the universe. We now have mathematical laws that govern everything we normally experience. It is a measure of our success that we now have to spend billions of dollars to build giant machines to accelerate particles to such high energy that we don't yet know what will happen when they collide. These very high particle energies don't occur in normal situations on earth, so it might seem academic and unnecessary to spend large sums on studying them. But they would have occurred in the early universe, so we must find out what happens at these energies if we are to understand how we and the universe began.

There is still a great deal that we don't know or understand about the universe. But the remarkable progress we have made, particularly in the last hundred years, should encourage us to believe that a complete understanding may not be beyond our powers. We may not be forever doomed to grope in the dark. We may break through to a complete theory of the universe. In that case, we would indeed be Masters of the Universe.

The scientific articles in this volume were written in the belief that the universe is governed by an order that we can perceive partially now and that we may understand fully in the not-too-distant future. It may be that this hope is just a mirage; there may be no ultimate theory, and even if there is, we may not find it. But it is surely better to strive for a complete understanding than to despair of the human mind.

STEPHEN HAWKING
31st March 1993

One

CHILDHOOD*

I WAS BORN ON January 8, 1942, exactly three hundred years after the death of Galileo. However, I estimate that about two hundred thousand other babies were also born that day. I don't know whether any of them were later interested in astronomy. I was born in Oxford, even though my parents were living in London. This was because Oxford was a good place to be born during World War II: The Germans had an agreement that they would not bomb Oxford and Cambridge, in re-

*This essay and the one that follows are based on a talk I gave to the International Motor Neurone Disease Society in Zurich in September 1987 and has been combined with material written in August 1991.

turn for the British not bombing Heidelberg and Göttingen. It is a pity that this civilized sort of arrangement couldn't have been extended to more cities.

My father came from Yorkshire. His grandfather, my great-grandfather, had been a wealthy farmer. He had bought too many farms and had gone bankrupt in the agricultural depression at the beginning of this century. This left my father's parents badly off, but they managed to send him to Oxford, where he studied medicine. He then went into research in tropical medicine. He went out to East Africa in 1937. When the war began, he made an overland journey across Africa to get a ship back to England, where he volunteered for military service. He was told, however, that he was more valuable in medical research.

My mother was born in Glasgow, Scotland, the second child of seven of a family doctor. The family moved south to Devon when she was twelve. Like my father's family, hers was not well off. Nevertheless, they managed to send my mother to Oxford. After Oxford, she had various jobs, including that of inspector of taxes, which she did not like. She gave that up to become a secretary. That was how she met my father in the early years of the war.

We lived in Highgate, north London. My sister Mary was born eighteen months after me. I'm told I did not welcome her arrival. All through our childhood there was a certain tension between us, fed by the narrow difference in our ages. In our adult life, however, this tension has disappeared, as we have gone different ways. She became a doctor, which pleased my father. My younger sister, Philippa, was born when I was nearly five and was able to understand what was happening. I can remember looking forward to her arrival so that there would be three of us to play games. She was a very intense and perceptive child. I always respected her judgment and opinions.

My brother Edward came much later, when I was fourteen, so he hardly entered my childhood at all. He was very different from the other three children, being completely nonacademic and nonintellectual. It was probably good for us. He was a rather difficult child, but one couldn't help liking him.

My earliest memory is of standing in the nursery of Byron House in Highgate and crying my head off. All around me, children were playing with what seemed like wonderful toys. I wanted to join in, but I was only two and a half, and this was the first time I had been left with people I didn't know. I think my parents were rather surprised at my reaction, because I was their first child and they had been following child development textbooks that said that children ought to start making social relationships at two. But they took me away after that awful morning and didn't send me back to Byron House for another year and a half.

At that time, during and just after the war, Highgate was an area in which a number of scientific and academic people lived. In another country they would have been called intellectuals, but the English have never admitted to having any intellectuals. All these parents sent their children to Byron House school, which was a very progressive school for those times. I remember complaining to my parents that they weren't teaching me anything. They didn't believe in what was then the accepted way of drilling things into you. Instead, you were supposed to learn to read without realizing you were being taught. In the end, I did learn to read, but not until the fairly late age of eight. My sister Philippa was taught to read by more conventional methods and could read by the age of four. But then, she was definitely brighter than me.

We lived in a tall, narrow Victorian house, which my parents had bought very cheaply during the war, when everyone thought London was going to be bombed flat. In fact, a V-2 rocket landed a few houses away from ours. I was away with my mother and sister at the time, but my father was in the house. Fortunately, he was not hurt, and the house was not badly damaged. But for years there was a large bomb site down the road, on which I used to play with my friend Howard, who lived three doors the other way. Howard was a revelation to me because his parents weren't intellectuals like the parents of all the other children I knew. He went to the council school, not Byron House, and he knew about football and boxing, sports that my parents wouldn't have dreamed of following.

Another early memory was getting my first train set. Toys were not manufactured during the war, at least not for the home market. But I had a passionate interest in model trains. My father tried making me a wooden train, but that didn't satisfy me, as I wanted something that worked. So my father got a secondhand clockwork train, repaired it with a soldering iron, and gave it to me for Christmas when I was nearly three. That train didn't work very well. But my father went to America just after the war, and when he came back on the *Queen Mary,* he brought my mother some nylons, which were not obtainable in Britain at that time. He brought my sister Mary a doll that closed its eyes when you laid it down. And he brought me an American train, complete with a cowcatcher and a figure-eight track. I can still remember my excitement as I opened the box.

Clockwork trains were all very well, but what I really wanted were electric trains. I used to spend hours watching a model railway club layout in Crouch End, near Highgate. I dreamed about electric trains. Finally, when both my parents were away somewhere, I took the opportunity to draw out of the Post Office bank all the very modest amount of money that people had given me on special occasions like my christening. I used the money to buy an electric train set, but frustratingly enough, it didn't work very well. Nowadays, we know about consumer rights. I should have taken the set back and demanded that the shop or manufacturer replace it, but in those days the attitude was that it was a privilege to buy something, and it was just your bad luck if it turned out to be faulty. So I paid for the electric motor of the engine to be serviced, but it never worked very well.

Later on, in my teens, I built model airplanes and boats. I was never very good with my hands, but I did this with my school friend John McClenahan, who was much better and whose father had a workshop in their house. My aim was always to build working models that I could control. I didn't care what they looked like. I think it was the same drive that led me to invent a series of very complicated games with another school friend, Roger Ferneyhough. There was a manufacturing game, complete with factories in which units of different colors were made, roads and railways on which they were carried, and a stock market. There was a war game, played on a board of four thousand squares, and even a feudal game, in which each player was a whole dynasty, with a family tree. I think these games, as well as the trains, boats, and airplanes, came from an urge to know how things worked and to control them. Since I began my Ph.D., this need has been met by my research into cosmology. If you understand how the universe operates, you control it in a way.

In 1950 my father's place of work moved from Hampstead, near Highgate, to the newly constructed National Institute for Medical Research in Mill Hill, on the northern edge of London. Rather than travel out from Highgate, it seemed more sensible to move out of London and travel in to town. My parents therefore bought a house in the cathedral city of St. Albans, about ten miles north of Mill Hill and twenty miles north of London. It was a large Victorian house of some elegance and character. My parents were not very well off when they bought it, and they had to have quite a lot of work done on it before we could move in. Thereafter my father, like the Yorkshireman he was, refused to pay for any further repairs. Instead, he did his best to keep it going and keep it painted, but it was a big house and he was not very skilled in such matters. The house was solidly built, however, so it withstood this neglect. My parents sold it in 1985, when my father was very ill (he died in 1986). I saw it recently. It didn't seem that any more work had been done on it, but it still looked much the same.

The house had been designed for a family with servants, and in the pantry there was an indicator board that showed which room the bell had been rung from. Of course we didn't have servants, but my first bedroom was a little L-shaped room that must have been a maid's room. I asked for it at the suggestion of my cousin Sarah, who was slightly older than me and whom I greatly admired. She said that we could have great fun there. One of the attractions of the room was that one could climb from the window out onto the roof of the bicycle shed and thence to the ground.

Sarah was the daughter of my mother's eldest sister, Janet, who had trained as a doctor and was married to a psychoanalyst. They lived in a rather similar house in Harpenden, a village five miles further north. They were one of the reasons we moved to St. Albans. It was a great bonus to me to be near Sarah, and I frequently went on the bus to Harpenden. St. Albans itself stood next to the remains of the ancient Roman city of Verulamium, which had been the most important Roman settlement in Britain after London. In the Middle Ages it had had the richest monastery in Britain. It was built around the shrine of Saint Alban, a Roman centurion who is said to be the first person in Britain to be executed for the Christian faith. All that remained of the abbey was a very large and rather ugly abbey church and the old abbey gateway building, which was now part of St. Albans school, where I later went.

St. Albans was a somewhat stodgy and conservative place compared with Highgate or Harpenden. My parents made hardly any friends there. In part this was their own fault, as they were naturally rather solitary, particularly my father. But it also reflected a different kind of population; certainly, none of the parents of my school friends in St. Albans could be described as intellectuals.

In Highgate our family had seemed fairly normal, but in St. Albans I think we were definitely regarded as eccentric. This perception was increased by the behavior of my father, who cared nothing for appearances if this allowed him to save money. His family had been very poor when he was young, and it had left a lasting impression on him. He couldn't bear to spend money on his own comfort, even when, in later years, he could afford to. He refused to put in central heating, even though he felt the cold badly. Instead, he would wear several sweaters and a dressing gown on top of his normal clothes. He was, however, very generous to other people.

In the 1950s he felt we couldn't afford a new car, so he bought a prewar London taxi, and he and I built a Nissen hut as a garage. The neighbors were outraged, but they couldn't stop us. Like most boys, I felt a need to conform, and I was embarrassed by my parents. But it never worried them.

When we first came to St. Albans, I was sent to the High School for Girls, which despite its name took boys up to the age of ten. After I had been there one term, however, my father took one of his almost yearly visits to Africa, this time for a rather longer period of about four months. My mother didn't feel like being left all that time, so she took my two sisters and me to visit her school friend Beryl, who was married to the poet Robert Graves. They lived in a village called Deya, on the Spanish island of Majorca. This was only five years after the war, and Spain's dictator, Francisco Franco, who had been an ally of Hitler and Mussolini, was still in power. (In fact, he remained in power for another two decades.) Nevertheless, my mother, who had been a member of the Young Communist League before the war, went with three young children by boat and train to Majorca. We rented a house in Deya and had a wonderful time. I shared a tutor with Robert's son, William. This tutor was a protégé of Robert and was more interested in writing a play for the Edinburgh festival than in teaching us. He therefore set us to read a chapter of the Bible each day and write a piece on it. The idea was to teach us the beauty of the English language. We got through all of Genesis and part of Exodus before I left. One of the main things I was taught from this was not to begin a sentence with *And*. I pointed out that most sentences in the Bible began with *And*, but I was told that English had changed since the time of King James. In that case, I argued, why make us read the Bible? But it was in vain. Robert Graves was very keen on the symbolism and mysticism in the Bible at that time.

When we got back from Majorca, I was sent to another school for a year, and then I took the so-called eleven-plus examination. This was an intelligence test that was taken at that time by all children who wanted state education. It has now been abolished, mainly because a number of middle-class children failed it and were sent to nonacademic schools. But I tended to do much better on tests and examinations than I did on coursework, so I passed the eleven-plus and got a free place at the local St. Albans school.

When I was thirteen my father wanted me to try for Westminster School, one of the main "public"—that is to say, private—schools. At that time there was a sharp division in education along class lines. My father felt that his lack of poise and connections had led him to being passed over in favor of people of less ability but more social graces. Because my parents were not well off, I would have to win a scholarship. I was ill at the time of the scholarship examination, however, and did not take it. Instead, I remained at St. Albans school. I got an education there that was as good as, if not better than, that I would have had at Westminster. I have never found that my lack of social graces has been a hindrance.

English education at that time was very hierarchical. Not only were schools divided into academic and nonacademic, but the academic schools were further divided into A, B, and C streams. This worked well for those in the A stream but not so well for those in the B stream, and badly for those in the C stream, who got discouraged. I was put in the A stream, based on the results of the eleven-plus. But after the first year, everyone who came below twentieth in the class was put down to the B stream. This was a tremendous blow to their self-confidence, from which some never recovered. In my first two terms at St. Albans, I came twenty-fourth and twenty-third, but in my third term I came eighteenth. So I just escaped.

I was never more than about halfway up the class. (It was a very bright class.) My classwork was very untidy, and my handwriting was the despair of my teachers. But my classmates gave me the nickname Einstein, so presumably they saw signs of something better. When I was twelve, one of my friends bet another friend a bag of sweets that I would never come to anything. I don't know if this bet was ever settled and, if so, which way it was decided.

I had six or seven close friends, most of whom I'm still in touch with. We used to have long discussions and arguments about everything from radio-controlled models to religion, and from parapsychology to physics. One of the things we talked about was the origin of the universe and whether it required a God to create it and set it going. I had heard that light from distant galaxies was shifted toward the red end of the spectrum and this was supposed to indicate that the universe was expanding. (A shift to the blue would have meant it was contracting.) But I was sure there must be some other reason for the red shift. Maybe light got tired, and more red, on its way to us. An essentially unchanging and everlasting universe seemed so much more natural. It was only after about two years of Ph.D. research that I realized I had been wrong.

When I came to the last two years of school, I wanted to specialize in mathematics and physics. There was an inspirational maths teacher, Mr. Tahta, and the school had just built a new maths room, which the maths set had as their classroom. But my father was very much against it. He thought there wouldn't be any jobs for mathematicians except as teachers. He would really have liked me to do medicine, but I showed no interest in biology, which seemed to me to be too descriptive and not sufficiently fundamental. It also had a rather low status at school. The brightest boys did mathematics and physics; the less bright did biology. My father knew I wouldn't do biology, but he made me do chemistry and only a small amount of mathematics. He felt this would keep my scientific options open. I'm now a professor of mathematics, but I have not had any formal instruction in mathematics since I left St. Albans school at the age of seventeen. I have had to pick up what mathematics I know as I went along. I used to supervise undergraduates at Cambridge and keep one week ahead of them in the course.

My father was engaged in research in tropical diseases, and he used to take me around his laboratory in Mill Hill. I enjoyed this, especially looking through microscopes. He also used to take me into the insect house, where he kept mosquitoes infected with tropical diseases. This worried me, because there always seemed to be a few mosquitoes flying around loose. He was very hard-working and dedicated to his research. He had a bit of a chip on his shoulder because he felt that other people who were not so good but who had the right background and connections had gotten ahead of him. He used to warn me against such people. But I think physics is a bit different from medicine. It doesn't matter what school you went to or to whom you are related. It matters what you do.

I was always very interested in how things operated and used to take them apart to see how they worked, but I was not so good at putting them back together again. My practical abilities never matched up to my theoretical inquiries. My father encouraged my interest in science, and he even coached me in mathematics until I got to a stage beyond his knowledge. With this background, and my father's job, I took it as natural that I would go into scientific research. In my early years I didn't differentiate between one kind of science and another. But from the age of thirteen or fourteen, I knew I wanted to do research in physics because it was the most fundamental science. This was despite the fact that physics was the most boring subject at school because it was so easy and obvious. Chemistry was much more fun because unexpected things, like explosions, kept happening. But physics and astronomy offered the hope of understanding where we came from and why we were here. I wanted to fathom the far depths of the universe. Maybe I have succeeded to a small extent, but there's still plenty I want to know.

Two

Oxford and Cambridge

M Y FATHER WAS very keen that I should go to Oxford or Cambridge. He himself had gone to University College, Oxford, so he thought I should apply there, because I would have a greater chance of getting in. At that time, University College had no fellow in mathematics, which was another reason he wanted me to do chemistry: I could try for a scholarship in natural science rather than in mathematics.

The rest of the family went to India for a year, but I had to stay behind to do A levels and university entrance. My headmaster thought I was much too young to try for Oxford, but I went up in March 1959 to do the scholarship exam with two boys from the year above me at school. I was convinced I had done badly and was very depressed when during the practical exam university lecturers came around to talk to other people but not to me. Then, a few days after I got back from Oxford, I got a telegram to say I had a scholarship.

I was seventeen, and most of the other students in my year had done military service and were a lot older. I felt rather lonely during my first year and part of the second. It was only in my third year that I really felt happy there. The prevailing attitude at Oxford at that time was very antiwork. You were supposed to be brilliant without effort, or to accept your limitations and get a fourth-class degree. To work hard to get a better class of degree was regarded as the mark of a gray man— the worst epithet in the Oxford vocabulary.

At that time, the physics course at Oxford was arranged in a way that made it particularly easy to avoid work. I did one exam before I went up, then had three years at Oxford with just the final exams at the end. I once calculated that I did about a thousand hours' work in the three years I was there, an average of an hour a day. I'm not proud of this lack of work. I'm just describing my attitude at the time, which I shared with most of my fellow students: an attitude of complete boredom and feeling that nothing was worth making an effort for. One result of my illness has been to change all that: When you are faced with the possibility of an early death, it makes you realize that life is worth living and that there are lots of things you want to do.

Because of my lack of work, I had planned to get through the final exam by doing problems in theoretical physics and avoiding questions that required factual knowledge. I didn't sleep the night before the exam because of nervous tension, however, so I didn't do very well. I was on the borderline between a first- and second-class degree, and I had to be interviewed by the examiners to determine which I should get. In the interview they asked me about my future plans. I replied that I wanted to do research. If they gave me a first, I would go to Cambridge. If I only got a second, I would stay in Oxford. They gave me a first.

I felt that there were two possible areas of theoretical physics that were fundamental and in which I might do research. One was cosmology, the study of the very large. The other was elementary particles, the study of the very small. I thought that elementary particles were less attractive because, although scientists were finding lots of new particles, there was no proper theory at that time. All they could do was arrange the particles in families, as in botany. In cosmology, on the other hand, there was a well-defined theory, Einstein's general theory of relativity.

There was then no one in Oxford working in cosmology, but at Cambridge there was Fred Hoyle, the most distinguished British astronomer of the time. I therefore applied to do a Ph.D. with Hoyle. My application to do research at Cambridge was accepted, provided I got a first, but to my annoyance my supervisor was not Hoyle but a man called Denis Sciama, of whom I had not heard. In the end, however, this turned out to be for the best. Hoyle was away abroad a lot, and I probably wouldn't have seen much of him. On the other hand, Sciama was there, and he was always stimulating, even though I often didn't agree with his ideas.

Because I had not done much mathematics at school or at Oxford, I found general relativity very difficult at first and did not make much progress. Also, during my last year at Oxford, I had noticed that I was getting rather clumsy in my movements. Soon after I went to Cambridge, I was diagnosed as having ALS, amyotrophic lateral sclerosis, or motor neurone disease, as it is known in England. (In the United States it is also called Lou

Gehrig's disease.) The doctors could offer no cure or assurance that it would not get worse.

At first the disease seemed to progress fairly rapidly. There did not seem much point in working at my research, because I didn't expect to live long enough to finish my Ph.D. As time went by, however, the disease seemed to slow down. I also began to understand general relativity and to make progress with my work. But what really made the difference was that I got engaged to a girl called Jane Wilde, whom I had met about the time I was diagnosed with ALS. This gave me something to live for.

If we were to get married, I had to get a job, and to get a job I had to finish my Ph.D. I therefore started working for the first time in my life. To my surprise, I found I liked it. Maybe it is not fair to call it work. Someone once said: Scientists and prostitutes get paid for doing what they enjoy.

I applied for a research fellowship at Gonville and Caius College (pronounced *Keys*). I was hoping that Jane would type my application, but when she came to visit me in Cambridge, she had her arm in plaster, having broken it. I must admit that I was less sympathetic than I should have been. It was her left arm, however, so she was able to write out the application to my dictation, and I got someone else to type it.

In my application I had to give the names of two people who could give references about my work. My supervisor suggested I should ask Hermann Bondi to be one of them. Bondi was then a professor of mathematics at Kings College, London, and an expert on general relativity. I had met him a couple of times, and he had submitted a paper I had written for publication in the journal *Proceedings of the Royal Society*. I asked him after a lecture he gave in Cambridge, and he looked at me in a vague way and said yes, he would. Obviously he didn't remember me, for when the College wrote to him for a reference, he replied that he had not heard of me. Nowadays, there are so many people applying for college research fellowships that if one of the candidate's referees says that he does not know him, that is the end of his chances. But those were quieter times. The College wrote to tell me of the embarrassing reply of my referee, and my supervisor got on to Bondi and refreshed his memory. Bondi then wrote me a reference that was probably far better than I deserved. I got the fellowship and have been a fellow of Caius College ever since.

The fellowship meant Jane and I could get married, which we did in July 1965. We spent a week's honeymoon in Suffolk, which was all I could afford. We then went to a summer school in general relativity at Cornell University in upstate New York. That was a mistake. We stayed in a dormitory that was full of couples with noisy small children, and it put quite a strain on our marriage. In other respects, however, the summer school was very useful for me because I met many of the leading people in the field.

My research up to 1970 was in cosmology, the study of the universe on a large scale. My most important work in this period was on singularities. Observations of distant galaxies indicate that they are moving away from us: The universe is expanding. This implies that the galaxies must have been closer together in the past. The question then arises: Was there a time in the past when all the galaxies were on top of each other and the density of the universe was infinite? Or was there a previous contracting phase, in which the galaxies managed to avoid hitting each other? Maybe they flew past each other and started to move away from each other. To answer this question required new mathematical techniques. These were developed between 1965 and 1970, mainly by Roger Penrose and myself. Penrose was then at Birkbeck College, London; now he is at Oxford. We used these techniques to show that there must have been a state of infinite density in the past, if the general theory of relativity is correct.

This state of infinite density is called the big bang singularity. It means that science would not be able to predict how the universe would begin, if general relativity is correct. However, my more recent work indicates that it is possible to predict how the universe would begin if one took into account the theory of quantum physics, the theory of the very small.

General relativity also predicts that massive stars will collapse in on themselves when they have exhausted their nuclear fuel. The work that Penrose and I did showed that they would continue to collapse until they reached a singularity of infinite density. This singularity would be an end of time, at least for the star and anything on it. The gravitational field of the singularity would be so strong that light could not escape from the region around it but would be dragged back by the gravitational field. The region from which it is not possible to escape is called a black hole, and its boundary is called the event horizon. Anything or anyone who falls into the black hole through the event horizon will come to an end of time at the singularity.

I was thinking about black holes as I got into bed one night in 1970, shortly after the birth of my daughter Lucy. Suddenly I realized that many of the techniques that Penrose and I had developed to prove singularities could be applied to black holes. In particular, the area of the event horizon, the boundary of the black hole, could not decrease with time. And when two black holes collided and joined together to form a single hole, the area of the horizon of the final hole would be greater than the sum of the areas of the horizons of the original black holes. This placed an important limit on the amount of energy that could be emitted in the collision. I was so excited that I did not get much sleep that night.

From 1970 to 1974 I worked mainly on black holes. But in

1974, I made perhaps my most surprising discovery: Black holes are not completely black! When one takes the small-scale behavior of matter into account, particles and radiation can leak out of a black hole. A black hole emits radiation as if it were a hot body.

Since 1974, I have been working on combining general relativity and quantum mechanics into a consistent theory. One result of that has been a proposal I made in 1983 with Jim Hartle of the University of California at Santa Barbara: that both time and space are finite in extent, but they don't have any boundary or edge. They would be like the surface of the earth, but with two more dimensions. The earth's surface is finite in area, but it doesn't have any boundary. In all my travels, I have not managed to fall off the edge of the world. If this proposal is correct, there would be no singularities, and the laws of science would hold everywhere, including at the beginning of the universe. The way the universe would begin would be determined by the laws of science. I would have succeeded in my ambition to discover *how* the universe began. But I still don't know *why* it began.

Three

MY EXPERIENCE
WITH ALS*

I AM QUITE OFTEN asked: How do you feel about having ALS? The answer is, not a lot. I try to lead as normal a life as possible and not think about my condition or regret the things it prevents me from doing, which are not that many.

It was a very great shock to me to discover that I had motor neurone disease. I had never been very well coordinated physically as a child. I was not good at ball games, and maybe for this reason I didn't care much for sport or physical activities. But things seemed to change when I went to Oxford. I took up

---

*A talk given to the British Motor Neurone Disease Association conference in Birmingham in October 1987.

21

coxing and rowing. I was not Boat Race standard, but I got by at the level of intercollege competition.

In my third year at Oxford, however, I noticed that I seemed to be getting clumsier, and I fell over once or twice for no apparent reason. But it was not until I was at Cambridge, in the following year, that my mother noticed and took me to the family doctor. He referred me to a specialist, and shortly after my twenty-first birthday I went into hospital for tests. I was in for two weeks, during which I had a wide variety of tests. They took a muscle sample from my arm, stuck electrodes into me, injected some radio-opaque fluid into my spine, and watched it going up and down with X-rays as they tilted the bed. After all that, they didn't tell me what I had, except that it was not multiple sclerosis and that I was an atypical case. I gathered, however, that they expected it to continue to get worse and that there was nothing they could do except give me vitamins. I could see that they didn't expect them to have much effect. I didn't feel like asking for more details, because they were obviously bad.

The realization that I had an incurable disease that was likely to kill me in a few years was a bit of a shock. How could something like that happen to me? Why should I be cut off like this? However, while I was in hospital, I had seen a boy I vaguely knew die of leukemia in the bed opposite me. It had not been a pretty sight. Clearly there were people who were worse off than me. At least my condition didn't make me feel sick. Whenever I feel inclined to be sorry for myself, I remember that boy.

Not knowing what was going to happen to me or how rapidly the disease would progress, I was at a loose end. The doctors told me to go back to Cambridge and carry on with the research I had just started in general relativity and cosmology. But I was not making much progress because I didn't have much mathematical background—and anyway, I might not live long enough to finish my Ph.D. I felt somewhat of a tragic character. I took to listening to Wagner, but reports in magazine articles that I drank heavily are an exaggeration. The trouble is, once one article said it, then other articles copied it because it made a good story. Anything that has appeared in print so many times must be true.

My dreams at that time were rather disturbed. Before my condition was diagnosed, I had been very bored with life. There had not seemed to be anything worth doing. But shortly after I came out of hospital, I dreamt that I was going to be executed. I suddenly realized that there were a lot of worthwhile things I could do if I were reprieved. Another dream that I had several times was that I would sacrifice my life to save others. After all, if I was going to die anyway, it might as well do some good.

But I didn't die. In fact, although there was a cloud hanging over my future, I found to my surprise that I was enjoying life in the present more than I had before. I began to make progress with my research, I got engaged and married, and I got a research fellowship at Caius College, Cambridge.

The fellowship at Caius took care of my immediate employment problem. I was lucky to have chosen to work in theoretical physics because that was one of the few areas in which my condition would not be a serious handicap. And I was fortunate that my scientific reputation increased at the same time that my disability got worse. This meant that people were prepared to offer me a sequence of positions in which I only had to do research without having to lecture.

We were also fortunate in housing. When we were married, Jane was still an undergraduate at Westfield College in London, so she had to go up to London during the week. This meant that we had to find somewhere I could manage on my own and that was centrally located, because I could not walk far. I asked the College if they could help, but was told by the then-bursar: It is College policy not to help fellows with housing. We therefore put our name down to rent one of a group of new flats that were being built in the marketplace. (Years later, I discovered that those flats were actually owned by the College, but they didn't tell me that.) When we returned to Cambridge from the summer in America, however, we found that the flats were not ready. As a great concession, the bursar offered us a room in a hostel for graduate students. He said, "We normally charge twelve shillings and sixpence a night for this room. However, as there will be two of you in the room, we will charge twenty-five shillings."

We stayed there only three nights. Then we found a small house about one hundred yards from my university department. It belonged to another College, which had let it to one of its fellows. He had recently moved out to a house in the suburbs, and he sublet the house to us for the remaining three months on his lease. During those three months, we found another house in the same road standing empty. A neighbor summoned the owner from Dorset and told her it was a scandal that her house should be vacant when young people were looking for accommodation, so she let the house to us. After we had lived there for a few years, we wanted to buy it and do it up, so we asked my College for a mortgage. The College did a survey and decided it was not a good risk. So in the end we got a mortgage from a building society, and my parents gave us the money to do it up.

We lived there for another four years, until it became too difficult for me to manage the stairs. By this time, the College appreciated me rather more and there was a different bursar. They therefore offered us a ground-floor flat in a house that they owned. This suited me very well because it had large rooms and wide doors. It was sufficiently central that I could get to my university department or the College in my electric wheelchair. It was also nice for our three children, because it was surrounded by a garden that was looked after by the College gardeners.

Up to 1974, I was able to feed myself and get in and out of bed. Jane managed to help me and bring up two children without outside help. Thereafter, however, things became more difficult, so we took to having one of my research students living with us. In return for free accommodation and a lot of my attention, they helped me get up and go to bed. In 1980 we changed to a system of community and private nurses who came in for an hour or two in the morning and evening. This lasted until I caught pneumonia in 1985. I had to have a tracheostomy operation, and from then on I needed twenty-four-hour nursing care. This was made possible by grants from several foundations.

Before the operation my speech had been getting more slurred, so that only people who knew me well could understand me. But at least I could communicate. I wrote scientific papers by dictating to a secretary, and I gave seminars through an interpreter who repeated my words more clearly. However, the tracheostomy removed my ability to speak altogether. For a time, the only way I could communicate was to spell out words letter by letter by raising my eyebrows when someone pointed to the right letter on a spelling card. It is pretty difficult to carry on a conversation like that, let alone write a scientific paper. However, a computer expert in California named Walt Woltosz heard of my plight. He sent me a computer program he had written called Equalizer. This allowed me to select words from a series of menus on the screen by pressing a switch in my hand. The program could also be controlled by a head or eye movement. When I have built up what I want to say, I can send it to a speech synthesizer.

At first, I just ran the Equalizer program on a desktop computer. Then David Mason, of Cambridge Adaptive Communications, fitted a small personal computer and a speech synthesizer to my wheelchair. This system allows me to communicate much better than I could before. I can manage up to fifteen words a minute. I can either speak what I have written or save it on disk. I can then print it out or call it back and speak it sentence by sentence. Using this system I have written two books and a number of scientific papers. I have also given a number of scientific and popular talks. They have been well received. I think that is in a large part due to the quality of the speech synthesizer, which is made by Speech Plus. One's voice is very important. If you have a slurred voice, people are likely to treat you as mentally deficient. This synthesizer is by far the best I have heard because it varies the intonation and doesn't speak like a Dalek. The only trouble is that it gives me an American accent. However, by now I identify with its voice. I would not want to change even if I were offered a British-sounding voice. I would feel I had become a different person.

I have had motor neurone disease for practically all my adult life. Yet it has not prevented me from having a very attractive family and being successful in my work. This is thanks to the help I have received from my wife, my children, and a large number of other people and organizations. I have been lucky that my condition has progressed more slowly than is often the case. It shows that one need not lose hope.

Four

PUBLIC ATTITUDES TOWARD SCIENCE*

W HETHER WE LIKE it or not, the world we live in has changed a great deal in the last hundred years, and it is likely to change even more in the next hundred. Some people would like to stop these changes and go back to what they see as a purer and simpler age. But as history shows, the past was not that wonderful. It was not so bad for a privileged minority, though even they had to do without modern medicine, and childbirth was highly risky for women. But for the

*A speech given in Oviedo, Spain, on receiving the Prince of Asturias Harmony and Concord Prize in October 1989. It has been updated.

vast majority of the population, life was nasty, brutish, and short.

Anyway, even if one wanted to, one couldn't put the clock back to an earlier age. Knowledge and techniques can't just be forgotten. Nor can one prevent further advances in the future. Even if all government money for research were cut off (and the present government is doing its best), the force of competition would still bring about advances in technology. Moreover, one cannot stop inquiring minds from thinking about basic science, whether or not they are paid for it. The only way to prevent further developments would be a global totalitarian state that suppressed anything new, and human initiative and ingenuity are such that even this wouldn't succeed. All it would do is slow down the rate of change.

If we accept that we cannot prevent science and technology from changing our world, we can at least try to ensure that the changes they make are in the right directions. In a democratic society, this means that the public needs to have a basic understanding of science, so that it can make informed decisions and not leave them in the hands of experts. At the moment, the public has a rather ambivalent attitude toward science. It has come to expect the steady increase in the standard of living that new developments in science and technology have brought to continue, but it also distrusts science because it doesn't understand it. This distrust is evident in the cartoon figure of the mad scientist working in his laboratory to produce a Frankenstein. It is also an important element behind support for the Green parties. But the public also has a great interest in science, particularly astronomy, as is shown by the large audiences for television series such as *Cosmos* and for science fiction.

What can be done to harness this interest and give the

public the scientific background it needs to make informed decisions on subjects like acid rain, the greenhouse effect, nuclear weapons, and genetic engineering? Clearly, the basis must lie in what is taught in schools. But in schools science is often presented in a dry and uninteresting manner. Children learn it by rote to pass examinations, and they don't see its relevance to the world around them. Moreover, science is often taught in terms of equations. Although equations are a concise and accurate way of describing mathematical ideas, they frighten most people. When I wrote a popular book recently, I was advised that each equation I included would halve the sales. I included one equation, Einstein's famous equation, $E = mc^2$. Maybe I would have sold twice as many copies without it.

Scientists and engineers tend to express their ideas in the form of equations because they need to know the precise values of quantities. But for the rest of us, a qualitative grasp of scientific concepts is sufficient, and this can be conveyed by words and diagrams, without the use of equations.

The science people learn in school can provide the basic framework. But the rate of scientific progress is now so rapid that there are always new developments that have occurred since one was at school or university. I never learned about molecular biology or transistors at school, but genetic engineering and computers are two of the developments most likely to change the way we live in the future. Popular books and magazine articles about science can help to put across new developments, but even the most successful popular book is read by only a small proportion of the population. Only television can reach a truly mass audience. There are some very good science programs on TV, but others present scientific wonders simply as magic, without explaining them or showing how they fit into the framework of scientific ideas. Producers of television science programs should realize that they have a responsibility to educate the public, not just entertain it.

What are the science-related issues that the public will have to make decisions on in the near future? By far the most urgent is that of nuclear weapons. Other global problems, such as food supply or the greenhouse effect, are relatively slow-acting, but a nuclear war could mean the end of all human life on earth within days. The relaxation of east-west tensions brought about by the ending of the cold war has meant that the fear of nuclear war has receded from public consciousness. But the danger is still there as long as there are enough weapons to kill the entire population of the world many times over. In former Soviet states and in America, nuclear weapons are still poised to strike all the major cities in the Northern Hemisphere. It would only take a computer error or a mutiny by some of those manning the weapons to trigger a global war. It is even more worrying that relatively minor powers are now acquiring nuclear weapons. The major powers have behaved in a reasonably responsible way, but one cannot have such confidence in minor powers like Libya or Iraq, Pakistan, or even Azerbaijan. The danger is not so much in the actual nuclear weapons that such powers may soon possess, which would be fairly rudimentary, though they could still kill millions of people. Rather, the danger is that a nuclear war between two minor powers could draw in the major powers with their enormous arsenals.

It is very important that the public realize the danger and put pressure on all governments to agree to large arms cuts. It probably is not practical to remove nuclear weapons entirely, but we can lessen the danger by reducing the number of weapons.

If we manage to avoid a nuclear war, there are still other dangers that could destroy us all. There's a sick joke that the reason we have not been contacted by an alien civilization is that civilizations tend to destroy themselves when they reach our stage. But I have sufficient faith in the good sense of the public to believe that we might prove this wrong.

Five

A BRIEF
HISTORY OF
*A BRIEF HISTORY**

I AM STILL RATHER taken aback by the reception given to my book, *A Brief History of Time*. It has been on *The New York Times* best-seller list for thirty-seven weeks and on *The Sunday Times* of London list for twenty-eight weeks. (It was published later in Britain than in the United States.) It is being translated into twenty languages (twenty-one if you count American as different from English). This was much more than

*This essay was originally published in December 1988 as an article in *The Independent*. *A Brief History of Time* remained on *The New York Times* best-seller list for fifty-three weeks; and in Britain, as of February 1993, it had been on *The Sunday Times* of London list for 205 weeks. (At week 184, it went into the *Guinness Book of Records* for achieving the most appearances on this list.) The number of translated editions is now thirty-three.

I expected when I first had the idea in 1982 of writing a popular book about the universe. My intention was partly to earn money to pay my daughter's school fees. (In fact, by the time the book actually appeared, she was in her last year of school.) But the main reason was that I wanted to explain how far I felt we had come in our understanding of the universe: how we might be near finding a complete theory that would describe the universe and everything in it.

If I were going to spend the time and effort to write a book, I wanted it to get to as many people as possible. My previous technical books had been published by Cambridge University Press. That publisher had done a good job, but I didn't feel that it would really be geared to the sort of mass market that I wanted to reach. I therefore contacted a literary agent, Al Zuckerman, who had been introduced to me as the brother-in-law of a colleague. I gave him a draft of the first chapter and explained that I wanted it to be the sort of book that would sell in airport book stalls. He told me there was no chance of that. It might sell well to academics and students, but a book like that couldn't break into Jeffrey Archer territory.

I gave Zuckerman a first draft of the book in 1984. He sent it to several publishers and recommended that I accept an offer from Norton, a fairly up-market American book firm. But I decided instead to take an offer from Bantam Books, a publisher more oriented toward the popular market. Though Bantam had not specialized in publishing science books, their books were widely available in airport book stalls. That they accepted my book was probably because of the interest in it taken by one of their editors, Peter Guzzardi. He took his job very seriously and made me rewrite the book to make it understandable to nonscientists like himself. Each time I sent him a rewritten chapter, he sent back a long list of objections and questions he wanted me to clarify. At times I thought the process would never end. But he was right: It is a much better book as a result.

Shortly after I accepted Bantam's offer, I got pneumonia. I had to have a tracheostomy operation that removed my voice. For a time I could communicate only by raising my eyebrows when someone pointed to letters on a card. It would have been quite impossible to finish the book but for the computer program I had been given. It was a bit slow, but then I think slowly, so it suited me quite well. With it I almost completely rewrote my first draft in response to Guzzardi's urgings. I was helped in this revision by one of my students, Brian Whitt.

I had been very impressed by Jacob Bronowski's television series, *The Ascent of Man.* (Such a sexist title would not be allowed today.) It gave a feeling for the achievement of the human race in developing from primitive savages only fifteen thousand years ago to our present state. I wanted to convey a similar feeling for our progress toward a complete understanding of the laws that govern the universe. I was sure that nearly everyone was interested in how the universe operates, but most people cannot follow mathematical equations—I don't care much for equations myself. This is partly because it is difficult for me to write them down but mainly because I don't have an intuitive feeling for equations. Instead, I think in pictorial terms, and my aim in the book was to describe these mental images in words, with the help of familiar analogies and a few diagrams. In this way, I hoped that most people would be able to share in the excitement and feeling of achievement in the remarkable progress that has been made in physics in the last twenty-five years.

Still, even if one avoids mathematics, some of the ideas are unfamiliar and difficult to explain. This posed a problem: Should I try to explain them and risk people being confused, or should I gloss over the difficulties? Some unfamiliar concepts, such as the fact that observers moving at different velocities measure different time intervals between the same pair of events, were not essential to the picture I wanted to draw. Therefore I felt I could just mention them but not go into depth. But other difficult ideas were basic to what I wanted to get across. There were two such concepts in particular that I felt I had to include. One was the so-called sum over histories. This is the idea that there is not just a single history for the universe. Rather, there is a collection of every possible history for the universe, and all these histories are equally real (whatever that may mean). The other idea, which is necessary to make mathematical sense of the sum over histories, is "imaginary time." With hindsight, I now feel that I should have put more effort into explaining these two very difficult concepts, particularly imaginary time, which seems to be the thing in the book with which people have the most trouble. However, it is not really necessary to understand exactly what imaginary time is—just that it is different from what we call real time.

When the book was nearing publication, a scientist who was sent an advance copy to review for *Nature* magazine was appalled to find it full of errors, with misplaced and erroneously labeled photographs and diagrams. He called Bantam, who were equally appalled and decided that same day to recall and scrap the entire printing. They spent three intense weeks correcting and rechecking the entire book, and it was ready in time to be in the bookstores by the April publication date. By then, *Time* magazine had published a profile of me. Even so, the editors were taken by surprise by the demand. The book is in its seventeenth printing in America and its tenth in Britain.*

Why did so many people buy it? It is difficult for me

---

*By April 1993, it was in its fortieth hardcover and nineteenth paperback printing in the United States, and its thirty-ninth hardcover printing in Britain.

to be sure that I'm objective, so I think I will go by what other people said. I found most of the reviews, although favorable, rather unilluminating. They tended to follow the formula: Stephen Hawking has Lou Gehrig's disease (in American reviews), or motor neurone disease (in British reviews). He is confined to a wheelchair, cannot speak, and can only move x number of fingers (where x seems to vary from one to three, according to which inaccurate article the reviewer read about me). Yet he has written this book about the biggest question of all: Where did we come from and where are we going? The answer that Hawking proposes is that the universe is neither created nor destroyed: It just *is*. In order to formulate this idea, Hawking introduces the concept of imaginary time, which I (the reviewer) find a little hard to follow. Still, if Hawking is right and we do find a complete unified theory, we shall really know the mind of God. (In the proof stage I nearly cut the last sentence in the book, which was that we would know the mind of God. Had I done so, the sales might have been halved.)

Rather more perceptive (I felt) was an article in *The Independent,* a London newspaper, which said that even a serious scientific book like *A Brief History of Time* could become a cult book. My wife was horrified, but I was rather flattered to have my book compared to *Zen and the Art of Motorcycle Maintenance.* I hope, like *Zen,* that it gives people the feeling that they need not be cut off from the great intellectual and philosophical questions.

Undoubtedly, the human interest story of how I have managed to be a theoretical physicist despite my disability has helped. But those who bought the book from the human interest angle may have been disappointed because it contains only a couple of references to my condition. The book was intended as a history of the universe, not of me. This has not prevented accusations that Bantam shamefully exploited my illness and that I cooperated with this by allowing my picture to appear on the cover. In fact, under my contract I had no control over the cover. I did, however, manage to persuade Bantam to use a better photograph on the British edition than the miserable and out-of-date photo used on the American edition. Bantam will not change the American cover, however, because it says that the American public now identifies that with the book.

It has also been suggested that people buy the book because they have read reviews of it or because it is on the bestseller list, but they don't read it; they just have it in the bookcase or on the coffee table, thereby getting credit for having it without taking the effort of having to understand it. I am sure this happens, but I don't know that it is any more so than for most other serious books, including the Bible and Shakespeare. On the other hand, I know that at least some people must have read it because each day I get a pile of letters about my book, many asking questions or making detailed comments that indicate that they have read it, even if they do not understand all of it. I also get stopped by strangers on the street who tell me how much they enjoyed it. Of course, I am more easily identified and more distinctive, if not distinguished, than most authors. But the frequency with which I receive such public congratulations (to the great embarrassment of my nine-year-old son) seems to indicate that at least a proportion of those who buy the book actually do read it.

People now ask me what I am going to do next. I feel I can hardly write a sequel to *A Brief History of Time*. What would I call it? *A Longer History of Time? Beyond the End of Time? Son of Time?* My agent has suggested that I allow a film to be made about my life. But neither I nor my family would have any self-respect left if we let ourselves be portrayed by actors. The same would be true to a lesser extent if I allowed and helped someone to write my life. Of course, I cannot stop someone from writing my life independently, as long as it is not libelous, but I try to put them off by saying I'm considering writing my autobiography. Maybe I will. But I'm in no hurry. I have a lot of science that I want to do first.

Six

MY POSITION*

T HIS ARTICLE IS not about whether I believe in God. In- stead, I will discuss my approach to how one can un- derstand the universe: what is the status and meaning of a grand unified theory, a "theory of everything." There is a real problem here. The people who ought to study and argue such questions, the philosophers, have mostly not had enough mathematical background to keep up with modern developments in theoret- ical physics. There is a subspecies called philosophers of sci- ence who ought to be better equipped. But many of them are failed physicists who found it too hard to invent new theories

*Originally given as a talk to a Caius College audience in May 1992.

41

and so took to writing about the philosophy of physics instead. They are still arguing about the scientific theories of the early years of this century, like relativity and quantum mechanics. They are not in touch with the present frontier of physics.

Maybe I'm being a bit harsh on philosophers, but they have not been very kind to me. My approach has been described as naive and simpleminded. I have been variously called a nominalist, an instrumentalist, a positivist, a realist, and several other ists. The technique seems to be refutation by denigration: If you can attach a label to my approach, you don't have to say what is wrong with it. Surely everyone knows the fatal errors of all those isms.

The people who actually make the advances in theoretical physics don't think in the categories that the philosophers and historians of science subsequently invent for them. I am sure that Einstein, Heisenberg, and Dirac didn't worry about whether they were realists or instrumentalists. They were simply concerned that the existing theories didn't fit together. In theoretical physics, the search for logical self-consistency has always been more important in making advances than experimental results. Otherwise elegant and beautiful theories have been rejected because they don't agree with observation, but I don't know of any major theory that has been advanced just on the basis of experiment. The theory always came first, put forward from the desire to have an elegant and consistent mathematical model. The theory then makes predictions, which can then be tested by observation. If the observations agree with the predictions, that doesn't prove the theory; but the theory survives to make further predictions, which again are tested against observation. If the observations don't agree with the predictions, one abandons the theory.

Or rather, that is what is supposed to happen. In practice, people are very reluctant to give up a theory in which they have invested a lot of time and effort. They usually start by questioning the accuracy of the observations. If that fails, they try to modify the theory in an ad hoc manner. Eventually the theory becomes a creaking and ugly edifice. Then someone suggests a new theory, in which all the awkward observations are explained in an elegant and natural manner. An example of this was the Michelson-Morley experiment, performed in 1887, which showed that the speed of light was always the same, no matter how the source or the observer was moving. This seemed ridiculous. Surely someone moving toward the light ought to measure it traveling at a higher speed than someone moving in the same direction as the light; yet the experiment showed that both observers would measure exactly the same speed. For the next eighteen years people like Hendrik Lorentz and George Fitzgerald tried to accommodate this observation within accepted ideas of space and time. They introduced ad hoc postulates, such as proposing that objects got shorter when they moved at high speeds. The entire framework of physics became clumsy and ugly. Then in 1905 Einstein suggested a much more attractive viewpoint, in which time was not regarded as completely separate and on its own. Instead it was combined with space in a four-dimensional object called space-time. Einstein was driven to this idea not so much by the experimental results as by the desire to make two parts of the theory fit together in a consistent whole. The two parts were the laws that govern the electric and magnetic fields, and the laws that govern the motion of bodies.

I don't think Einstein, or anyone else in 1905, realized how simple and elegant the new theory of relativity was. It completely revolutionized our notions of space and time. This example illustrates well the difficulty of being a realist in the philosophy of science, for what we regard as reality is conditioned by the theory to which we subscribe. I am certain Lorentz and Fitzgerald regarded themselves as realists, interpreting the experiment on the speed of light in terms of Newtonian ideas of absolute space and absolute time. These notions of space and time seemed to correspond to common sense and reality. Yet nowadays those who are familiar with the theory of relativity, still a disturbingly small minority, have a rather different view. We ought to be telling people about the modern understanding of such basic concepts as space and time.

If what we regard as real depends on our theory, how can we make reality the basis of our philosophy? I would say that I am a realist in the sense that I think there is a universe out there waiting to be investigated and understood. I regard the solipsist position that everything is the creation of our imaginations as a waste of time. No one acts on that basis. But we cannot distinguish what is real about the universe without a theory. I therefore take the view, which has been described as simple-minded or naive, that a theory of physics is just a mathematical model that we use to describe the results of observations. A theory is a good theory if it is an elegant model, if it describes a wide class of observations, and if it predicts the results of new observations. Beyond that, it makes no sense to ask if it corresponds to reality, because we do not know what reality is independent of a theory. This view of scientific theories may make me an instrumentalist or a positivist—as I have said above, I have been called both. The person who called me a positivist went on to add that everyone knew that positivism was out of date—another case of refutation by denigration. It may indeed be out of date in that it was yesterday's intellectual fad, but the positivist position I have outlined seems the only possible one for someone who is seeking new laws, and new ways, to describe the universe. It is no good appealing to reality because we don't have a model independent concept of reality.

In my opinion, the unspoken belief in a model independent reality is the underlying reason for the difficulties philosophers of science have with quantum mechanics and the uncertainty principle. There is a famous thought experiment called Schrödinger's cat. A cat is placed in a sealed box. There is a gun pointing at it, and it will go off if a radioactive nucleus decays. The probability of this happening is fifty percent. (Today no one would dare propose such a thing, even purely as a thought experiment, but in Schrödinger's time they had not heard of animal liberation.)

If one opens the box, one will find the cat either dead or alive. But before the box is opened, the quantum state of the cat will be a mixture of the dead cat state with a state in which the cat is alive. This some philosophers of science find very hard to accept. The cat can't be half shot and half not-shot, they claim, any more than one can be half pregnant. Their difficulty arises because they are implicitly using a classical concept of reality in which an object has a definite single history. The whole point of quantum mechanics is that it has a different view of reality. In this view, an object has not just a single history but all possible histories. In most cases, the probability of having a particular history will cancel out with the probability of having a very slightly different history; but in certain cases, the probabilities of neighboring histories reinforce each other. It is one of these reinforced histories that we observe as the history of the object.

In the case of Schrödinger's cat, there are two histories that are reinforced. In one the cat is shot, while in the other it remains alive. In quantum theory both possibilities can exist together. But some philosophers get themselves tied in knots because they implicitly assume that the cat can have only one history.

The nature of time is another example of an area in which our theories of physics determine our concept of reality. It used to be considered obvious that time flowed on forever, regardless of what was happening; but the theory of relativity combined time with space and said that both could be warped, or distorted, by the matter and energy in the universe. So our perception of the nature of time changed from being independent of the universe to being shaped by it. It then became conceivable that time might simply not be defined before a certain point; as one goes back in time, one might come to an insurmountable barrier, a singularity, beyond which one could not go. If that were the case, it wouldn't make sense to ask who, or what, caused or created the big bang. To talk about causation or creation implicitly assumes there was a time before the big bang singularity. We have known for twenty-five years that Einstein's general theory of relativity predicts that time must have had a beginning in a singularity fifteen billion years ago. But the philosophers have not yet caught up with the idea. They are still worrying about the foundations of quantum mechanics that were laid down sixty-five years ago. They don't realize that the frontier of physics has moved on.

Even worse is the mathematical concept of imaginary time, in which Jim Hartle and I suggested the universe may not have any beginning or end. I was savagely attacked by a philosopher of science for talking about imaginary time. He said: How can a mathematical trick like imaginary time have anything to do with the real universe? I think this philosopher was confusing the technical mathematical terms real and imaginary numbers with the way that real and imaginary are used in everyday language. This just illustrates my point: How can we know what is real, independent of a theory or model with which to interpret it?

I have used examples from relativity and quantum mechanics to show the problems one faces when one tries to make sense of the universe. It doesn't really matter if you don't understand relativity and quantum mechanics, or even if these theories are incorrect. What I hope I have demonstrated is that some sort of positivist approach, in which one regards a theory as a model, is the only way to understand the universe, at least for a theoretical physicist. I am hopeful that we will find a consistent model that describes everything in the universe. If we do that, it will be a real triumph for the human race.

Seven

IS THE END
IN SIGHT FOR
THEORETICAL
PHYSICS?*

I N THESE PAGES I want to discuss the possibility that the
goal of theoretical physics might be achieved in the not-
too-distant future: say, by the end of the century. By this I mean
that we might have a complete, consistent, and unified theory
of the physical interactions that would describe all possible observations. Of course, one has to be very cautious about making
such predictions. We have thought that we were on the brink
of the final synthesis at least twice before. At the beginning of
the century it was believed that everything could be understood

*On April 29, 1980 I was inaugurated as Lucasian Professor of Mathematics at Cam-
bridge. This essay, my Inaugural Lecture, was read for me by one of my students.

in terms of continuum mechanics. All that was needed was to measure a certain number of coefficients of elasticity, viscosity, conductivity, etc. This hope was shattered by the discovery of atomic structure and quantum mechanics. Again, in the late 1920s Max Born told a group of scientists visiting Göttingen that "physics, as we know it, will be over in six months." This was shortly after the discovery by Paul Dirac, a previous holder of the Lucasian Chair, of the Dirac equation, which governs the behavior of the electron. It was expected that a similar equation would govern the proton, the only other supposedly elementary particle known at that time. However, the discovery of the neutron and of nuclear forces disappointed those hopes. We now know in fact that neither the proton nor the neutron is elementary but that they are made up of smaller particles. Nevertheless, we have made a lot of progress in recent years, and as I shall describe, there are some grounds for cautious optimism that we may see a complete theory within the lifetime of some of those reading these pages.

Even if we do achieve a complete unified theory, we shall not be able to make detailed predictions in any but the simplest situations. For example, we already know the physical laws that govern everything that we experience in everyday life. As Dirac pointed out, his equation was the basis of "most of physics and all of chemistry." However, we have been able to solve the equation only for the very simplest system, the hydrogen atom, consisting of one proton and one electron. For more complicated atoms with more electrons, let alone for molecules with more than one nucleus, we have to resort to approximations and intuitive guesses of doubtful validity. For macroscopic systems consisting of 10^{23} particles or so, we have to use statistical methods and abandon any pretense of solving the equations exactly. Although in principle we know the equations that govern the whole of biology, we have not been able to reduce the study of human behavior to a branch of applied mathematics.

What would we mean by a complete and unified theory of physics? Our attempts at modeling physical reality normally consist of two parts:

1. A set of local laws that are obeyed by the various physical quantities. These are usually formulated in terms of differential equations.
2. Sets of boundary conditions that tell us the state of some regions of the universe at a certain time and what effects propagate into it subsequently from the rest of the universe.

Many people would claim that the role of science is confined to the first of these and that theoretical physics will have achieved its goal when we have obtained a complete set of local physical laws. They would regard the question of the initial conditions for the universe as belonging to the realm of metaphysics or religion. In a way, this attitude is similar to that of those who in earlier centuries discouraged scientific investigation by saying that all natural phenomena were the work of God and should not be inquired into. I think that the initial conditions of the universe are as suitable a subject for scientific study and theory as are the local physical laws. We shall not have a complete theory until we can do more than merely say that "things are as they are because they were as they were."

The question of the uniqueness of the initial conditions is closely related to that of the arbitrariness of the local physical laws: One would not regard a theory as complete if it contained a number of adjustable parameters such as masses or coupling constants that could be given any values one liked. In fact, it seems that neither the initial conditions nor the values of the parameters in the theory are arbitrary but that they are somehow chosen or picked out very carefully. For example, if the proton-neutron mass difference were not about twice the mass of the electron, one would not obtain the couple of hundred or so stable nucleides that make up the elements and are the basis of chemistry and biology. Similarly, if the gravitational mass of the proton were significantly different, one would not have had stars in which these nucleides could have been built up, and if the initial expansion of the universe had been slightly smaller or slightly greater, the universe would either have collapsed before such stars could have evolved or would have expanded so rapidly that stars would never have been formed by gravitational condensation.

Indeed, some people have gone so far as to elevate these restrictions on the initial conditions and the parameters to the status of a principle, the anthropic principle, which can be paraphrased as, "Things are as they are because we are." According to one version of the principle, there is a very large number of different, separate universes with different values of the physical parameters and different initial conditions. Most of these universes will not provide the right conditions for the development of the complicated structures needed for intelligent life. Only in a small number, with conditions and parameters like our own universe, will it be possible for intelligent life to develop and to ask the question, "Why is the universe as we observe it?" The answer, of course, is that if it were otherwise, there would not be anyone to ask the question.

The anthropic principle does provide some sort of explanation of many of the remarkable numerical relations that are observed between the values of different physical parameters. However, it is not completely satisfactory; one cannot help feeling that there is some deeper explanation. Also, it cannot account for all the regions of the universe. For example, our solar system is certainly a prerequisite for our existence, as is an earlier generation of nearby stars in which heavy elements could have been formed by nuclear synthesis. It might even be that the whole of our galaxy was required. But there does not seem any necessity for other galaxies to exist, let alone the million million or so of them that we see distributed roughly uniformly throughout the observable universe. This large-scale homogeneity of the universe makes it very difficult to believe that the structure of the universe is determined by anything so peripheral as some complicated molecular structures on a minor planet orbiting a very average star in the outer suburbs of a fairly typical spiral galaxy.

If we are not going to appeal to the anthropic principle, we need some unifying theory to account for the initial conditions of the universe and the values of the various physical parameters. However, it is too difficult to think up a complete theory of everything all at one go (though this does not seem to stop some people; I get two or three unified theories in the mail each week). What we do instead is to look for partial theories that will describe situations in which certain interactions can be ignored or approximated in a simple manner. We first divide the material content of the universe into two parts: "matter," particles such as quarks, electrons, muons, etc., and "interactions," such as gravity, electromagnetism, etc. The matter particles are described by fields of one-half-integer spin and obey the Pauli exclusion principle, which prevents more than one particle of a given kind from being in any state. This is the reason we can have solid bodies that do not collapse to a point or radiate away to infinity. The matter principles are divided into two groups: the hadrons, which are composed of quarks; and the leptons, which comprise the remainder.

The interactions are divided phenomenologically into four categories. In order of strength, they are: the strong nuclear forces, which interact only with hadrons; electromagnetism, which interacts with charged hadrons and leptons; the weak nuclear forces, which interact with all hadrons and leptons; and finally, the weakest by far, gravity, which interacts with everything. The interactions are represented by integer-spin fields that do not obey the Pauli exclusion principle. This means they can have many particles in the same state. In the case of electromagnetism and gravity, the interactions are also long-range, which means that the fields produced by a large number of matter particles can all add up to give a field that can be detected on a macroscopic scale. For these reasons, they were the first to have theories developed for them: gravity by Newton in the seventeenth century, and electromagnetism by Maxwell in the nineteenth century. However, these theories were basically incompatible because the Newtonian theory was invariant if the whole system was given any uniform velocity, whereas the Maxwell theory defined a preferred velocity—the speed of light. In the end, it turned out to be the Newtonian theory of gravity that had to be modified to make it compatible with the invariance properties of the Maxwell theory. This was achieved by Einstein's general theory of relativity, which was formulated in 1915.

The general relativity theory of gravity and the Maxwell theory of electrodynamics were what are called classical theories; that is, they involved quantities that were continuously variable and that could, in principle at least, be measured to arbitrary accuracy. However, a problem arose when one tried to use such theories to construct a model of the atom. It had been discovered that the atom consisted of a small, positively charged nucleus surrounded by a cloud of negatively charged electrons. The natural assumption was that the electrons were in orbit around the nucleus as the earth is in orbit around the sun. But the classical theory predicted that the electrons would radiate electromagnetic waves. These waves would carry away energy and would cause the electrons to spiral into the nucleus, producing the collapse of the atom.

This problem was overcome by what is undoubtedly the greatest achievement in theoretical physics in this century: the discovery of the quantum theory. The basic postulate of this is the Heisenberg uncertainty principle, which states that certain pairs of quantities, such as the position and momentum of a particle, cannot be measured simultaneously with arbitrary accuracy. In the case of the atom, this meant that in its lowest energy state the electron could not be at rest in the nucleus because, in that case, its position would be exactly defined (at the nucleus) and its velocity would also be exactly defined (to be zero). Instead, both position and velocity would have to be smeared out with some probability distribution around the nucleus. In this state the electron could not radiate energy in the form of electromagnetic waves because there would be no lower energy state for it to go to.

In the 1920s and 1930s quantum mechanics was applied with great success to systems such as atoms or molecules, which have only a finite number of degrees of freedom. Difficulties arose, however, when people tried to apply it to the electromagnetic field, which has an infinite number of degrees of freedom, roughly speaking two for each point of space-time. One can regard these degrees of freedom as oscillators, each with its own position and momentum. The oscillators cannot be at rest because then they would have exactly defined positions and momenta. Instead, each oscillator must have some minimum amount of what are called zero-point fluctuations and a nonzero energy. The energies of all the infinite number of degrees of freedom would cause the apparent mass and charge of the electron to become infinite.

A procedure called renormalization was developed to overcome this difficulty in the late 1940s. It consisted of the rather arbitrary subtraction of certain infinite quantities to leave finite remainders. In the case of electrodynamics, it was necessary to make two such infinite subtractions, one for the mass and the other for the charge of the electron. This renormalization procedure has never been put on a very firm conceptual or mathematical basis, but it has worked quite well in practice. Its great success was the prediction of a small displacement, the Lamb shift, in some lines in the spectrum of atomic hydrogen. However, it is not very satisfactory from the point of view of attempts to construct a complete theory because it does not make any predictions of the values of the finite remainders left after making infinite subtractions. Thus, we would have to fall back on the anthropic principle to explain why the electron has the mass and charge that it does.

During the 1950s and 1960s it was generally believed that the weak and strong nuclear forces were not renormalizable; that is, they would require an infinite number of infinite subtractions to make them finite. There would be an infinite number of finite remainders that were not determined by the theory. Such a theory would have no predictive power because one could never measure all the infinite number of parameters. However, in 1971 Gerard 't Hooft showed that a unified model of the electromagnetic and weak interactions that had been proposed earlier by Abdus Salam and Steven Weinberg was indeed renormalizable with only a finite number of infinite subtractions. In the Salam-Weinberg theory the photon, the spin-1 particle that carries the electromagnetic interaction, is joined by three other spin-1 partners called W^+, W^-, and Z^0. At very high energies these four particles are all predicted to behave in a similar manner. However, at lower energies a phenomenon called spontaneous symmetry breaking is invoked to explain the fact that the photon has zero rest mass, whereas the W^+, W^-, and Z^0 are all very massive. The low-energy predictions of this theory have agreed remarkably well with observation, and this led the Swedish Academy in 1979 to award the Nobel Prize in physics to Salam, Weinberg, and Sheldon Glashow, who had also constructed similar unified theories. However, Glashow himself remarked that the Nobel committee really took rather a gamble, because we do not yet have particle accelerators of high enough energy to test the theory in the regime where unification between the electromagnetic forces, carried by the photon, and the weak forces, carried by the W^+, W^-, and Z^0, really occurs. Sufficiently powerful accelerators will be ready in a few years, and most physicists are confident that they will confirm the Salam-Weinberg theory.*

The success of the Salam-Weinberg theory led to the search for a similar renormalizable theory of the strong interactions. It was realized fairly early on that the proton and other hadrons such as the pi meson could not be truly elementary particles, but that they must be bound states of other particles called quarks. These seem to have the curious property that, although they can move fairly freely within a hadron, it appears to be impossible to obtain just one quark on its own; they always come either in groups of three (like the proton or neutron) or in pairs consisting of a quark and antiquark (like the pi meson). To explain this, quarks were endowed with an attribute called color. It should be emphasized that this has nothing to do with our normal perception of color; quarks are far too small to be seen by visible light. It is merely a convenient name. The idea is that quarks come in three colors—red, green, and blue— but that any isolated bound state, such as a hadron, has to be colorless, either a combination of red, green, and blue like the

*In fact, the W and Z particles were observed at the CERN laboratory in Geneva in 1983 and another Nobel Prize was awarded in 1984 to Carlo Rubbia and Simon van der Meere, who led the team that made the discovery. The person who missed out on a prize was 't Hooft.

proton, or a mixture of red and antired, green and antigreen, and blue and antiblue, like the pi meson.

The strong interactions between the quarks are supposed to be carried by spin-1 particles called gluons, rather like the particles that carry the weak interaction. The gluons also carry color, and they and the quarks obey a renormalizable theory called quantum chromodynamics, or QCD for short. A consequence of the renormalization procedure is that the effective coupling constant of the theory depends on the energy at which it is measured and decreases to zero at very high energies. This phenomenon is known as asymptotic freedom. It means that quarks inside a hadron behave almost like free particles in high-energy collisions, so that their perturbations can be treated successfully by perturbation theory. The predictions of perturbation theory are in reasonable qualitative agreement with observation, but one cannot yet really claim that the theory has been experimentally verified. At low energies the effective coupling constant becomes very large and perturbation theory breaks down. It is hoped that this "infrared slavery" will explain why quarks are always confined in colorless bound states, but so far no one has been able to demonstrate this really convincingly.

Having obtained one renormalizable theory for the strong interactions and another one for the weak and electromagnetic interactions, it was natural to look for a theory that combined the two. Such theories are given the rather exaggerated title "grand unified theories," or GUTs. This is rather misleading because they are neither all that grand, nor fully unified, nor complete theories in that they have a number of undetermined renormalization parameters such as coupling constants and masses. Nevertheless, they may be a significant step toward a complete unified theory. The basic idea is that the effective coupling constant of the strong interactions, which is large at low energies, gradually decreases at high energies because of asymptotic freedom. On the other hand, the effective coupling constant of the Salam-Weinberg theory, which is small at low energies, gradually increases at high energies because this theory is not asymptotically free. If one extrapolates the low-energy rate of increase and decrease of the coupling constants, one finds that the two coupling constants become equal at an energy of about 10^{15} GeV. (GeV means a billion electron volts. This is about the energy that would be released if a hydrogen atom could be totally converted into energy. By comparison, the energy released in chemical reactions like burning is on the order of one electron volt per atom.) The theories propose that above this energy the strong interactions are unified with the weak and electromagnetic interactions, but that at lower energies there is spontaneous symmetry breaking.

An energy of 10^{15} GeV is way beyond the scope of any laboratory equipment; the present generation of particle accelerators can produce center-of-mass energies of about 10 GeV, and the next generation will produce energies of 100 GeV or so. This will be sufficient to investigate the energy range in which the electromagnetic forces should become unified with the weak forces according to the Salam-Weinberg theory, but not the enormously high energy at which the weak and electromagnetic interactions would be predicted to become unified with the strong interactions. Nevertheless, there can be low-energy predictions of the grand unified theories that might be testable in the laboratory. For example, the theories predict that the proton should not be completely stable but should decay with a lifetime of order 10^{31} years. The present experimental lower limit on the lifetime is about 10^{30} years, and it should be possible to improve this.

Another observable prediction concerns the ratio of baryons to photons in the universe. The laws of physics seem to be the same for particles and antiparticles. More precisely, they are the same if particles are replaced by antiparticles, right-handed is replaced by left-handed, and the velocities of all particles are reversed. This is known as the CPT theorem, and it is a consequence of basic assumptions that should hold in any reasonable theory. Yet the earth, and indeed the whole solar system, is made up of protons and neutrons without any antiprotons or antineutrons. Indeed, such an imbalance between particles and antiparticles is yet another a priori condition for our existence, for if the solar system were composed of an equal mixture of particles and antiparticles, they would all annihilate each other and leave just radiation. From the observed absence of such annihilation radiation we can conclude that our galaxy is made entirely of particles rather than antiparticles. We do not have direct evidence of other galaxies, but it seems likely that they are composed of particles and that in the universe as a whole, there is an excess of particles over antiparticles of about one particle per 10^8 photons. One could try to account for this by invoking the anthropic principle, but grand unified theories actually provide a possible mechanism for explaining the imbalance. Although all interactions seem to be invariant under the combination of C (replace particles by antiparticles), P (change right-handed to left-handed), and T (reverse the direction of time), there are known to be interactions that are not invariant under T alone. In the early universe, in which there is a very marked arrow of time given by the expansion, these interactions could produce more particles than antiparticles. However, the number they make is very model-dependent, so that agreement with observation is hardly a confirmation of the grand unified theories.

So far, most of the effort has been devoted to unifying the first three categories of physical interactions, the strong and weak nuclear forces and electromagnetism. The fourth and last, gravity, has been neglected. One justification for this is that gravity is so weak that quantum gravitational effects would be large only at particle energies way beyond those in any particle accelerator. Another is that gravity does not seem to be renormalizable; in order to obtain finite answers, it seems that one may have to make an infinite number of infinite subtractions with a correspondingly infinite number of undetermined finite remainders. Yet one must include gravity if one is to obtain a fully unified theory. Furthermore, the classical theory of general relativity predicts that there should be space-time singularities at which the gravitational field would become infinitely strong. These singularities would occur in the past at the beginning of the present expansion of the universe (the big bang), and in the future in the gravitational collapse of stars and, possibly, of the universe itself. The prediction of singularities presumably indicates that the classical theory will break down. However, there seems to be no reason why it should break down until the gravitational field becomes strong enough that quantum gravitational effects are important. Thus, a quantum theory of gravity is essential if we are to describe the early universe and then give some explanation for the initial conditions beyond merely appealing to the anthropic principle.

Such a theory is also required if we are to answer the question: Does time really have a beginning and, possibly, an end, as is predicted by classical general relativity, or are the singularities in the big bang and the big crunch smeared out in some way by quantum effects? This is a difficult question to give a well-defined meaning to when the very structures of space and time themselves are subject to the uncertainty principle. My personal feeling is that singularities are probably still present, though one can continue time past them in a certain mathematical sense. However, any subjective concept of time that was related to consciousness or the ability to perform measurements would come to an end.

What are the prospects of obtaining a quantum theory of gravity and of unifying it with the other three categories of interactions? The best hope seems to lie in an extension of general relativity called supergravity. In this, the graviton, the spin-2 particle that carries the gravitational interaction, is related to a number of other fields of lower spin by so-called supersymmetry transformations. Such a theory has the great merit that it does away with the old dichotomy between "matter," represented by particles of one-half-integer spin, and "interactions," represented by integer-spin particles. It also has the great advantage that many of the infinities that arise in quantum theory cancel each other out. Whether they all cancel out to give a theory that is finite without any infinite subtractions is not yet known. It is hoped that they do, because it can be shown that theories that include gravity are either finite or nonrenormalizable; that is, if one has to make any infinite subtractions, then one will have to make an infinite number of them with a corresponding infinite number of undetermined remainders. Thus, if all the infinities in supergravity turn out to cancel each other out, we could have a theory that not only fully unifies all the matter particles and interactions, but that is complete in the sense that it does not have any undetermined renormalization parameters.

Although we do not yet have a proper quantum theory of gravity, let alone one that unifies it with the other physical interactions, we do have an idea of some of the features it should have. One of these is connected with the fact that gravity affects the causal structure of space-time; that is, gravity determines which events can be causally related to one another. An example of this in the classical theory of general relativity is provided by a black hole, which is a region of space-time in which the gravitational field is so strong that any light or other signal is dragged back into the region and cannot escape to the outside world. The intense gravitational field near the black hole causes the creation of pairs of particles and antiparticles, one of which falls into the black hole and the other of which escapes to infinity. The particle that escapes appears to have been emitted by the black hole. An observer at a distance from the black hole can measure only the outgoing particles, and he cannot correlate them with those that fall into the hole because he cannot observe them. This means that the outgoing particles have an extra degree of randomness or unpredictability over and above that usually associated with the uncertainty principle. In normal situations the uncertainty principle implies that one can definitely predict *either* the position *or* the velocity of a particle *or* one combination of position and velocity. Thus, roughly speaking, one's ability to make definite predictions is halved. However, in the case of particles emitted from a black hole, the fact that one cannot observe what is going on inside the black hole means that one can definitely predict *neither* the positions *nor* the velocities of the emitted particles. All one can give are probabilities that particles will be emitted in certain modes.

It seems, therefore, that even if we find a unified theory, we may be able to make only statistical predictions. We would also have to abandon the view that there is a unique universe that we observe. Instead, we would have to adopt a picture in which there is an ensemble of all possible universes with some probability distribution. This might explain why the universe started off in the big bang in almost perfect thermal equilibrium, because thermal equilibrium would correspond to the largest number of microscopic configurations and hence the greatest probability. To paraphrase Voltaire's philosopher, Pangloss, "We live in the most probable of all possible worlds."

What are the prospects that we will find a complete unified theory in the not-too-distant future? Each time we have extended our observations to smaller length scales and higher energies, we have discovered new layers of structure. At the beginning of the century, the discovery of Brownian motion with a typical energy particle of 3×10^{-2} eV showed that matter is not continuous but is made up of atoms. Shortly thereafter, it was discovered that these supposedly indivisible atoms are made up of electrons revolving about a nucleus with energies of the order of a few electron-volts. The nucleus, in turn, was found to be composed of so-called elementary particles, protons and neutrons, held together by nuclear bonds of the order of 10^6 eV. The latest episode in this story is that we have found that the proton and electron are made up of quarks held together by bonds of the order of 10^9 eV. It is a tribute to how far we have come already in theoretical physics that it now takes enormous machines and a great deal of money to perform an experiment whose results we cannot predict.

Our past experience might suggest that there is an infinite sequence of layers of structure at higher and higher energies. Indeed, such a view of an infinite regress of boxes within boxes was official dogma in China under the Gang of Four. However, it seems that gravity should provide a limit, but only at the very short length scale of 10^{-33} cm or the very high energy of 10^{28} eV. On length scales shorter than this, one would expect that space-time would cease to behave like a smooth continuum and that it would acquire a foamlike structure because of quantum fluctuations of the gravitational field.

There is a very large unexplored region between our present experimental limit of about 10^{10} eV and the gravitational cutoff at 10^{28} eV. It might seem naive to assume, as grand unified theories do, that there are only one or two layers of structure in this enormous interval. However, there are grounds for optimism. At the moment, at least, it seems that gravity can be unified with the other physical interactions only in some supergravity theory. There appears to be only a finite number of such theories. In particular, there is a largest such theory, the so-called $N = 8$ extended supergravity. This contains one graviton, eight spin-$\frac{3}{2}$ particles called gravitonos, twenty-eight spin-1 particles, fifty-six spin-$\frac{1}{2}$ particles, and seventy particles of spin 0. Large as these numbers are, they are not large enough to account for all the particles that we seem to observe in strong and weak interactions. For instance, the $N = 8$ theory has twenty-eight spin-1 particles. These are sufficient to account for the gluons that carry the strong interactions and two of the four particles that carry the weak interactions, but not the other two. One would therefore have to believe that many or most of the observed particles such as gluons or quarks are not really elementary, as they seem at the moment, but that they are bound states of the fundamental $N = 8$ particles. It is not likely that we shall have accelerators powerful enough to probe these composite structures within the foreseeable future, or indeed ever, if one makes a projection based on current economic trends. Nevertheless, the fact that these bound states arose from the well-defined $N = 8$ theory should enable us to make a number of predictions that could be tested at energies that are accessible now or will be in the near future. The situation might thus be similar to that for the Salam-Weinberg theory unifying electromagnetism and weak interactions. The low-energy predictions of this theory are in such good agreement with observation that the theory is now generally accepted, even though we have not yet reached the energy at which the unification should take place.

There ought to be something very distinctive about the theory that describes the universe. Why does this theory come to life while other theories exist only in the minds of their inventors? The $N = 8$ supergravity theory does have some claims to be special. It seems that it may be the only theory

1. that is in four dimensions
2. that incorporates gravity
3. that is finite without any infinite subtractions

I have already pointed out that the third property is necessary if we are to have a complete theory without parameters. It is, however, difficult to account for properties 1 and 2 without appealing to the anthropic principle. There seems to be a consistent theory that satisfies properties 1 and 3 but that does not include gravity. However, in such a universe there would probably not be sufficient in the way of attractive forces to gather together matter in the large aggregates that are probably necessary for the development of complicated structures. Why space-time should be four-dimensional is a question that is normally considered to be outside the realm of physics. However, there is a good anthropic principle argument for that too. Three space-time dimensions—that is, two space and one time—are clearly insufficient for any complicated organism. On the other hand, if there were more than three spatial dimensions, the orbits of planets around the sun or electrons around a nucleus would be unstable and they would tend to spiral inward. There remains the possibility of more than one time dimension, but I for one find such a universe very hard to imagine.

So far, I have implicitly assumed that there is an ultimate theory. But is there? There are at least three possibilities:

1. There is a complete unified theory.
2. There is no ultimate theory, but there is an infinite sequence of theories that are such that any particular class

of observations can be predicted by taking a theory sufficiently far down the chain.

3. There is no theory. Observations cannot be described
   or predicted beyond a certain point but are just arbitrary.

The third view was advanced as an argument against the
scientists of the seventeenth and eighteenth centuries: How
could they formulate laws that would curtail the freedom of
God to change His mind? Nevertheless they did, and they got
away with it. In modern times we have effectively eliminated
possibility 3 by incorporating it within our scheme: Quantum
mechanics is essentially a theory of what we do not know and
cannot predict.

Possibility 2 would amount to a picture of an infinite sequence of structures at higher and higher energies. As I said
before, this seems unlikely because one would expect that
there would be a cutoff at the Planck energy of 10^{28} eV. This
leaves us with possibility 1. At the moment the $N = 8$ supergravity theory is the only candidate in sight.* There are likely
to be a number of crucial calculations within the next few years
that have the possibility of showing that the theory is no good.
If the theory survives these tests, it will probably be some years
more before we develop computational methods that will enable us to make predictions and before we can account for the
initial conditions of the universe as well as the local physical
laws. These will be the outstanding problems for theoretical
physicists in the next twenty years or so. But to end on a slightly

*Supergravity theories seem to be the only particle theory with properties 1, 2, and 3,
but since this was written, there has been a great wave of interest in what are called
superstring theories. In these the basic objects are not point particles but extended
objects like little loops of string. The idea is that what appears to us to be a particle is
really a vibration on a loop. These superstring theories seem to reduce to supergravity
in the low-energy limit, but so far there has been little success in obtaining experimentally testable predictions from superstring theory.

alarmist note, they may not have much more time than that. At present, computers are a useful aid in research, but they have to be directed by human minds. If one extrapolates their recent rapid rate of development, however, it would seem quite possible that they will take over altogether in theoretical physics. So maybe the end is in sight for theoretical physicists, if not for theoretical physics.

Eight

EINSTEIN'S DREAM*

IN THE EARLY years of the twentieth century, two new theories completely changed the way we think about space and time, and about reality itself. More than seventy-five years later, we are still working out their implications and trying to combine them in a unified theory that will describe everything in the universe. The two theories are the general theory of relativity and quantum mechanics. The general theory of relativity deals with space and time and how they are curved or warped on a large scale by the matter and energy in the universe. Quan-

*A lecture given at the Paradigm Session of the NTT Data Communications Systems Corporation in Tokyo in July 1991.

tum mechanics, on the other hand, deals with very small scales. Included in it is what is called the uncertainty principle, which states that one can never precisely measure the position and the velocity of a particle at the same time; the more accurately you can measure one, the less accurately you can measure the other. There is always an element of uncertainty or chance, and this affects the behavior of matter on a small scale in a fundamental way. Einstein was almost singlehandedly responsible for general relativity, and he played an important part in the development of quantum mechanics. His feelings about the latter are summed up in the phrase "God does not play dice." But all the evidence indicates that God is an inveterate gambler and that He throws the dice on every possible occasion.

In this essay, I will try to convey the basic ideas behind these two theories, and why Einstein was so unhappy about quantum mechanics. I shall also describe some of the remarkable things that seem to happen when one tries to combine the two theories. These indicate that time itself had a beginning about fifteen billion years ago and that it may come to an end at some point in the future. Yet in another kind of time, the universe has no boundary. It is neither created nor destroyed. It just is.

I shall start with the theory of relativity. National laws hold only within one country, but the laws of physics are the same in Britain, the United States, and Japan. They are also the same on Mars and in the Andromeda galaxy. Not only that, the laws are the same at no matter what speed you are moving. The laws are the same on a bullet train or on a jet airplane as they are for someone standing in one place. In fact, of course, even someone who is stationary on the earth is moving at about 18.6 miles (30 kilometers) a second around the sun. The sun is also moving at several hundred kilometers a second around the galaxy, and so on. Yet all this motion makes no difference to the laws of physics; they are the same for all observers.

This independence of the speed of the system was first discovered by Galileo, who developed the laws of motion of objects like cannonballs or planets. However, a problem arose when people tried to extend this independence of the speed of the observer to the laws that govern the motion of light. It had been discovered in the eighteenth century that light does not travel instantaneously from source to observer; rather, it goes at a certain speed, about 186,000 miles (300,000 kilometers) a second. But what was this speed relative *to*? It seemed that there had to be some medium throughout space through which the light traveled. This medium was called the ether. The idea was that light waves traveled at a speed of 186,000 miles a second through the ether, which meant that an observer who was at rest relative to the ether would measure the speed of light to be about 186,000 miles a second, but an observer who was moving through the ether would measure a higher or lower speed. In particular, it was believed that the speed of light ought to change as the earth moves through the ether on its orbit around the sun. However, in 1887 a careful experiment carried out by Michelson and Morley showed that the speed of light was always the same. No matter what speed the observer was moving at, he would always measure the speed of light at 186,000 miles a second.

How can this be true? How can observers moving at different speeds all measure light at the same speed? The answer is they can't, not if our normal ideas of space and time hold true. However, in a famous paper written in 1905, Einstein pointed out that such observers could all measure the same speed of light if they abandoned the idea of a universal time. Instead, they would each have their own individual time, as measured by a clock each carried with him. The times measured by these different clocks would agree almost exactly if they were moving slowly with respect to each other—but the times measured by different clocks would differ significantly if the clocks were moving at high speed. This effect has actually been observed by comparing a clock on the ground with one in a commercial airliner; the clock in the airliner runs slightly slow when compared to the stationary clock. However, for normal speeds of travel, the differences between the rates of clocks are very small. You would have to fly around the world four hundred million times to add one second to your life; but your life would be reduced by more than that by all those airline meals.

How does having their own individual time cause people traveling at different speeds to measure the same speed of light? The speed of a pulse of light is the distance it travels between two events, divided by the time interval between the events. (An event in this sense is something that takes place at a single point in space, at a specified point in time.) People moving at different speeds will not agree on the distance between two events. For example, if I measure a car traveling down the highway, I might think it had moved only one kilometer, but to someone on the sun, it would have moved about 1,800 kilometers, because the earth would have moved while the car was going down the road. Because people moving at different speeds measure different distances between events, they must also measure different intervals of time if they are to agree on the speed of light.

Einstein's original theory of relativity, which he proposed in the paper written in 1905, is what we now call the special theory of relativity. It describes how objects move through space and time. It shows that time is not a universal quantity which exists on its own, separate from space. Rather, future and past are just directions, like up and down, left and right, forward and back, in something called space-time. You can only go in the future direction in time, but you *can* go at a bit of an angle to it. That is why time can pass at different rates.

The special theory of relativity combined time with space, but space and time were still a fixed background in which events happened. You could choose to move on different paths through space-time, but nothing you could do would modify the background of space and time. However, all this was changed when Einstein formulated the general theory of relativity in 1915. He had the revolutionary idea that gravity was not just a force that operated in a fixed background of space-time. Instead, gravity was a *distortion* of space-time, caused by the mass and energy in it. Objects like cannonballs and planets try to move on a straight line through space-time, but because space-time is curved, warped, rather than flat, their paths appear to be bent. The earth is trying to move on a straight line through space-time, but the curvature of space-time produced by the mass of the sun causes it to go in a circle around the sun. Similarly, light tries to travel in a straight line, but the curvature of space-time near the sun causes the light from distant stars to be bent if it passes near the sun. Normally, one is not able to see stars in the sky that are in almost the same direction as the sun. During an eclipse, however, when most of the sun's light is blocked off by the moon, one can observe the light from those stars. Einstein produced his general theory of relativity during the First World War, when conditions were not suitable for scientific observations, but immediately after the war a British expedition observed the eclipse of 1919 and confirmed the predictions of general relativity: Space-time is not flat, but is curved by the matter and energy in it.

This was Einstein's greatest triumph. His discovery completely transformed the way we think about space and time. They were no longer a passive background in which events took place. No longer could we think of space and time as running on forever, unaffected by what happened in the universe. Instead, they were now dynamic quantities that influenced and were influenced by events that took place in them.

An important property of mass and energy is that they are always positive. This is why gravity always attracts bodies toward each other. For example, the gravity of the earth attracts us to it even on opposite sides of the world. That is why people in Australia don't fall off the world. Similarly, the gravity of the sun keeps the planets in orbit around it and stops the earth from shooting off into the darkness of interstellar space. According to general relativity, the fact that mass is always positive means that space-time is curved back on itself, like the surface of the earth. If mass had been negative, space-time would have been curved the other way, like the surface of a saddle. This positive curvature of space-time, which reflects the fact that gravity is attractive, was seen as a great problem by Einstein. It was then widely believed that the universe was static, yet if space, and particularly time, were curved back on themselves, how could the universe continue forever in more or less the same state as it is at the present time?

Einstein's original equations of general relativity predicted that the universe was either expanding or contracting. Einstein therefore added a further term to the equations that relate the mass and energy in the universe to the curvature of space-time. This so-called cosmological term had a repulsive gravitational effect. It was thus possible to balance the attraction of the matter with the repulsion of the cosmological term. In other words, the negative curvature of space-time produced by the cosmological term could cancel the positive curvature of space-time produced by the mass and energy in the universe. In this way, one could obtain a model of the universe that continued forever in the same state. Had Einstein stuck to his original equations, without the cosmological term, he would have predicted that the universe was either expanding or contracting. As it was, no one thought the universe was changing with time until 1929, when Edwin Hubble discovered that distant galaxies are moving away from us. The universe is expanding. Einstein later called the cosmological term "the greatest mistake of my life."

But with or without the cosmological term, the fact that matter caused space-time to curve in on itself remained a problem, though it was not generally recognized as such. What it meant was that matter could curve a region in on itself so much that it would effectively cut itself off from the rest of the universe. The region would become what is called a black hole. Objects could fall into the black hole, but nothing could escape. To get out, they would need to travel faster than the speed of light, which is not allowed by the theory of relativity. Thus the matter inside the black hole would be trapped and would collapse to some unknown state of very high density.

Einstein was deeply disturbed by the implications of this collapse, and he refused to believe that it happened. But Robert Oppenheimer showed in 1939 that an old star of more than twice the mass of the sun would inevitably collapse when it had exhausted all its nuclear fuel. Then war intervened, Oppenheimer became involved in the atom bomb project, and he lost interest in gravitational collapse. Other scientists were more concerned with physics that could be studied on earth. They distrusted predictions about the far reaches of the universe because it did not seem they could be tested by observation. In the 1960s, however, the great improvement in the range and quality of astronomical observations led to new interest in gravitational collapse and in the early universe. Exactly what Einstein's general theory of relativity predicted in these situations remained unclear until Roger Penrose and I proved a number of theorems. These showed that the fact that space-time was curved in on itself implied that there would be singularities, places where space-time had a beginning or an end. It would have had a beginning in the big bang, about fifteen billion years ago, and it would come to an end for a star that collapsed and for anything that fell into the black hole the collapsing star left behind.

The fact that Einstein's general theory of relativity turned out to predict singularities led to a crisis in physics. The equations of general relativity, which relate the curvature of space-time with the distribution of mass and energy, cannot be defined as a singularity. This means that general relativity cannot predict what comes out of a singularity. In particular, general relativity cannot predict how the universe should begin at the big bang. Thus, general relativity is not a complete theory. It needs an added ingredient in order to determine how the universe should begin and what should happen when matter collapses under its own gravity.

The necessary extra ingredient seems to be quantum mechanics. In 1905, the same year he wrote his paper on the special theory of relativity, Einstein also wrote about a phenomenon called the photoelectric effect. It had been observed that when light fell on certain metals, charged particles were given off. The puzzling thing was that if the intensity of the light was reduced, the number of particles emitted diminished, but the speed with which each particle was emitted remained the same. Einstein showed this could be explained if light came not in continuously variable amounts, as everyone had assumed, but rather in packets of a certain size. The idea of light coming only in packets, called quanta, had been introduced a few years earlier by the German physicist Max Planck. It is a bit like saying one can't buy sugar loose in a supermarket but only in kilogram bags. Planck used the idea of quanta to explain why a red-hot piece of metal doesn't give off an infinite amount of heat; but he regarded quanta simply as a theoretical trick, one that didn't correspond to anything in physical reality.

Einstein's paper showed that you could directly observe individual quanta. Each particle emitted corresponded to one quantum of light hitting the metal. It was widely recognized to be a very important contribution to quantum theory, and it won him the Nobel Prize in 1922. (He should have won a Nobel Prize for general relativity, but the idea that space and time were curved was still regarded as too speculative and controversial, so they gave him a prize for the photoelectric effect instead—not that it was not worth the prize on its own account.)

The full implications of the photoelectric effect were not realized until 1925, when Werner Heisenberg pointed out that it made it impossible to measure the position of a particle exactly. To see where a particle is, you have to shine light on it. But Einstein had shown that you couldn't use a very small amount of light; you had to use at least one packet, or quantum. This packet of light would disturb the particle and cause it to move at a speed in some direction. The more accurately you wanted to measure the position of the particle, the greater the energy of the packet you would have to use and thus the more it would disturb the particle. However you tried to measure the particle, the uncertainty in its position, times the uncertainty in its speed, would always be greater than a certain minimum amount.

This uncertainty principle of Heisenberg showed that one could not measure the state of a system exactly, so one could not predict exactly what it would do in the future. All one could do is predict the probabilities of different outcomes. It was this element of chance, or randomness, that so disturbed Einstein. He refused to believe that physical laws should not make a definite, unambiguous prediction for what would happen. But however one expresses it, all the evidence is that the quantum phenomenon and the uncertainty principle are unavoidable and that they occur in every branch of physics.

Einstein's general relativity is what is called a classical theory; that is, it does not incorporate the uncertainty principle. One therefore has to find a new theory that combines general relativity with the uncertainty principle. In most situations, the difference between this new theory and classical general relativity will be very small. This is because, as noted earlier, the uncertainty predicted by quantum effects is only on very small scales, while general relativity deals with the structure of space-time on very large scales. However, the singularity theorems that Roger Penrose and I proved show that space-time will become highly curved on very small scales. The effects of the uncertainty principle will then become very important and seem to point to some remarkable results.

Part of Einstein's problems with quantum mechanics and the uncertainty principle arose from the fact that he used the ordinary, commonsense notion that a system has a definite history. A particle is either in one place or in another. It can't be half in one and half in another. Similarly, an event like the landing of astronauts on the moon either has taken place or it hasn't. It cannot have half-taken place. It's like the fact that you can't be slightly dead or slightly pregnant. You either are or you aren't. But if a system has a single definite history, the uncertainty principle leads to all sorts of paradoxes, like the particles being in two places at once or astronauts being only half on the moon.

An elegant way to avoid these paradoxes that had so troubled Einstein was put forward by the American physicist Richard Feynman. Feynman became well known in 1948 for work on the quantum theory of light. He was awarded the Nobel Prize in 1965 with another American, Julian Schwinger, and the Japanese physicist Shinichiro Tomonaga. But he was a physicist's physicist, in the same tradition as Einstein. He hated pomp and humbug, and he resigned from the National Academy of

Sciences because he found that they spent most of their time deciding which other scientists should be admitted to the Academy. Feynman, who died in 1988, is remembered for his many contributions to theoretical physics. One of these was the diagrams that bear his name, which are the basis of almost every calculation in particle physics. But an even more important contribution was his concept of a sum over histories. The idea was that a system didn't have just a single history in space-time, as one would normally assume it did in a classical nonquantum theory. Rather, it had every possible history. Consider, for example, a particle that is at a point A at a certain time. Normally, one would assume that the particle will move on a straight line away from A. However, according to the sum over histories, it can move on *any* path that starts at A. It is like what happens when you place a drop of ink on a piece of blotting paper. The particles of ink will spread through the blotting paper along every possible path. Even if you block the straight line between two points by putting a cut in the paper, the ink will get around the corner.

Associated with each path or history of the particle will be a number that depends on the shape of the path. The probability of the particle traveling from A to B is given by adding up the numbers associated with all the paths that take the particle from A to B. For most paths, the number associated with the path will nearly cancel out the numbers from paths that are close by. Thus, they will make little contribution to the probability of the particle's going from A to B. But the numbers from the straight paths will add up with the numbers from paths that are almost straight. Thus the main contribution to the probability will come from paths that are straight or almost straight. That is why the track a particle makes when going through a bubble chamber looks almost straight. But if you put something like a wall with a slit in it in the way of the particle, the particle paths can spread out beyond the slit. There can be a high probability of finding the particle away from the direct line through the slit.

In 1973 I began investigating what effect the uncertainty principle would have on a particle in the curved space-time near a black hole. Remarkably enough, I found that the black hole would not be completely black. The uncertainty principle would allow particles and radiation to leak out of the black hole at a steady rate. This result came as a complete surprise to me and everyone else, and it was greeted with general disbelief. But with hindsight, it ought to have been obvious. A black hole is a region of space from which it is impossible to escape if one is traveling at less than the speed of light. But the Feynman sum over histories says that particles can take *any* path through space-time. Thus it is possible for a particle to travel faster than light. The probability is low for it to move a long distance at more than the speed of light, but it can go faster than light for just far enough to get out of the black hole, and then go slower than light. In this way, the uncertainty principle allows particles to escape from what was thought to be the ultimate prison, a black hole. The probability of a particle getting out of a black hole of the mass of the sun would be very low because the particle would have to travel faster than light for several kilometers. But there might be very much smaller black holes, which were formed in the early universe. These primordial black holes could be less than the size of the nucleus of an atom, yet their mass could be a billion tons, the mass of Mount Fuji. They could be emitting as much energy as a large power station. If only we could find one of these little black holes and harness its energy! Unfortunately, there don't seem to be many around in the universe.

The prediction of radiation from black holes was the first nontrivial result of combining Einstein's general relativity with the quantum principle. It showed that gravitational collapse was not as much of a dead end as it had appeared to be. The particles in a black hole need not have an end of their histories at a singularity. Instead, they could escape from the black hole and continue their histories outside. Maybe the quantum principle would mean that one could also avoid the histories having a beginning in time, a point of creation, at the big bang.

This is a much more difficult question to answer, because it involves applying the quantum principle to the structure of time and space themselves and not just to particle paths in a given space-time background. What one needs is a way of doing the sum over histories not just for particles but for the whole fabric of space and time as well. We don't know yet how to do this summation properly, but we do know certain features it should have. One of these is that it is easier to do the sum if one deals with histories in what is called imaginary time rather than in ordinary, real time. Imaginary time is a difficult concept to grasp, and it is probably the one that has caused the greatest problems for readers of my book. I have also been criticized fiercely by philosophers for using imaginary time. How can imaginary time have anything to do with the real universe? I think these philosophers have not learned the lessons of history. It was once considered obvious that the earth was flat and that the sun went around the earth, yet since the time of Copernicus and Galileo, we have had to adjust to the idea that the earth is round and that it goes around the sun. Similarly, it was long obvious that time went at the same rate for every observer, but since Einstein, we have had to accept that time goes at different rates for different observers. It also seemed obvious that the universe had a unique history, yet since the discovery of quantum mechanics, we have had to consider the universe as having every possible history. I want to suggest that the idea of imaginary time is something that we will also have to come to accept. It is an intellectual leap of the same order as believing that the world is round. I think that imaginary time will come to seem as natural as a round earth does now. There are not many Flat Earthers left in the educated world.

You can think of ordinary, real time as a horizontal line, going from left to right. Early times are on the left, and late times are on the right. But you can also consider another direction of time, up and down the page. This is the so-called imaginary direction of time, at right angles to real time.

What is the point of introducing the concept of imaginary time? Why doesn't one just stick to the ordinary, real time that we understand? The reason is that, as noted earlier, matter and energy tend to make space-time curve in on itself. In the real time direction, this inevitably leads to singularities, places where space-time comes to an end. At the singularities, the equations of physics cannot be defined; thus one cannot predict what will happen. But the imaginary time direction is at right angles to real time. This means that it behaves in a similar way to the three directions that correspond to moving in space. The curvature of space-time caused by the matter in the universe can then lead to the three space directions and the imaginary time direction meeting up around the back. They would form a closed surface, like the surface of the earth. The three space directions and imaginary time would form a space-time that was closed in on itself, without boundaries or edges. It wouldn't have any point that could be called a beginning or end, any more than the surface of the earth has a beginning or end.

In 1983, Jim Hartle and I proposed that the sum over histories for the universe should not be taken over histories in real time. Rather, it should be taken over histories in imaginary time that were closed in on themselves, like the surface of the earth. Because these histories didn't have any singularities or any beginning or end, what happened in them would be determined entirely by the laws of physics. This means that what happened in imaginary time could be calculated. And if you know the history of the universe in imaginary time, you can calculate how it behaves in real time. In this way, you could hope to get a complete unified theory, one that would predict everything in the universe. Einstein spent the later years of his life looking for such a theory. He did not find one because he distrusted quantum mechanics. He was not prepared to admit that the universe could have many alternative histories, as in the sum over histories. We still do not know how to do the sum over histories properly for the universe, but we can be fairly sure that it will involve imaginary time and the idea of space-time closing up on itself. I think these concepts will come to seem as natural to the next generation as the idea that the world is round. Imaginary time is already a commonplace of science fiction. But it is more than science fiction or a mathematical trick. It is something that shapes the universe we live in.

Nine

THE ORIGIN
OF THE
UNIVERSE*

T HE PROBLEM OF the origin of the universe is a bit like the old question: Which came first, the chicken or the egg? In other words, what agency created the universe, and what created that agency? Or perhaps the universe, or the agency that created it, existed forever and didn't need to be created. Up to recently, scientists have tended to shy away from such questions, feeling that they belong to metaphysics or religion rather than to science. In the last few years, however, it

*A lecture given at the Three Hundred Years of Gravity conference held in Cambridge in June 1987, on the three hundredth anniversary of the publication of Newton's *Principia*.

has emerged that the laws of science may hold even at the beginning of the universe. In that case the universe could be self-contained and determined completely by the laws of science.

The debate about whether and how the universe began has been going on throughout recorded history. Basically, there were two schools of thought. Many early traditions, and the Jewish, Christian, and Islamic religions, held that the universe was created in the fairly recent past. (In the seventeenth century Bishop Ussher calculated a date of 4004 B.C. for the creation of the universe, a figure he arrived at by adding up the ages of people in the Old Testament.) One fact that was used to support the idea of a recent origin was the recognition that the human race is obviously evolving in culture and technology. We remember who first performed that deed or developed this technique. Thus, the argument runs, we cannot have been around all that long; otherwise, we would have already progressed more than we have. In fact, the biblical date for the creation is not that far off the date of the end of the last ice age, which is when modern humans seem first to have appeared.

On the other hand, there were people such as the Greek philosopher Aristotle who did not like the idea that the universe had a beginning. They felt that would imply divine intervention. They preferred to believe that the universe had existed and would exist forever. Something that was eternal was more perfect than something that had to be created. They had an answer to the argument about human progress described above: Periodic floods or other natural disasters had repeatedly set the human race right back to the beginning.

Both schools of thought held that the universe was

essentially unchanging with time. Either it was created in its present form, or it has endured forever as it is today. This was a natural belief, because human life—indeed, the whole of recorded history—is so brief that during it the universe has not changed significantly. In a static, unchanging universe, the question of whether it has existed forever or whether it was created at a finite time in the past is really a matter for metaphysics or religion: Either theory could account for such a universe. Indeed, in 1781 the philosopher Immanuel Kant wrote a monumental and very obscure work, *The Critique of Pure Reason,* in which he concluded that there were equally valid arguments both for believing that the universe had a beginning and for believing that it did not. As his title suggests, his conclusions were based simply on reason; in other words, they did not take any account of observations of the universe. After all, in an unchanging universe, what was there to observe?

In the nineteenth century, however, evidence began to accumulate that the earth and the rest of the universe were in fact changing with time. Geologists realized that the formation of the rocks and the fossils in them would have taken hundreds or thousands of millions of years. This was far longer than the age of the earth as calculated by the creationists. Further evidence was provided by the so-called second law of thermodynamics, formulated by the German physicist Ludwig Boltzmann. It states that the total amount of disorder in the universe (which is measured by a quantity called entropy) always increases with time. This, like the argument about human progress, suggests that the universe can have been going only for a finite time. Otherwise, it would by now have degenerated into a state of complete disorder, in which everything would be at the same temperature.

Another difficulty with the idea of a static universe was that according to Newton's law of gravity, each star in the universe ought to be attracted toward every other star. If so, how could they stay motionless, at a constant distance from each other? Wouldn't they all fall together?

Newton was aware of this problem. In a letter to Richard Bentley, a leading philosopher of the time, he agreed that a *finite* collection of stars could not remain motionless; they would all fall together to some central point. However, he argued, an infinite collection of stars would not fall together, for there would not be any central point for them to fall to. This argument is an example of the pitfalls that one can encounter when one talks about infinite systems. By using different ways to add up the forces on each star from the infinite number of other stars in the universe, one can get different answers to the question of whether the stars can remain at constant distances from each other. We now know that the correct procedure is to consider the case of a *finite* region of stars, and then to add more stars, distributed roughly uniformly outside the region. A finite collection of stars will fall together, and according to Newton's law, adding more stars outside the region will not stop the collapse. Thus, an infinite collection of stars cannot remain in a motionless state. If they are not moving relative to each other at one time, the attraction between them will cause them to start falling toward each other. Alternatively, they can be moving away from each other, with gravity slowing down the velocity of the recession.

Despite these difficulties with the idea of a static and unchanging universe, no one in the seventeenth, eighteenth, nineteenth, or early twentieth century suggested that the universe might be evolving with time. Newton and Einstein both missed the chance of predicting that the universe should be either contracting or expanding. One cannot really hold it against Newton, because he lived two hundred and fifty years before the observational discovery of the expansion of the universe. But Einstein should have known better. The theory of general relativity he formulated in 1915 predicted that the universe was expanding. But he remained so convinced of a static universe that he added an element to his theory to reconcile it with Newton's theory and balance gravity.

The discovery of the expansion of the universe by Edwin Hubble in 1929 completely changed the discussion about its origin. If you take the present notion of the galaxies and run it back in time, it would seem that they should all have been on top of each other at some moment between ten and twenty thousand million years ago. At this time, a singularity called the big bang, the density of the universe and the curvature of space-time would have been infinite. Under such conditions, all the known laws of science would break down. This is a disaster for science. It would mean that science alone could not predict how the universe began. All that science could say is: The universe is as it is now because it was as it was then. But science could not explain why it was as it was just after the big bang.

Not surprisingly, many scientists were unhappy with this conclusion. There were thus several attempts to avoid the conclusion that there must have been a big bang singularity and hence a beginning of time. One was the so-called steady state theory. The idea was that, as the galaxies moved apart from each other, new galaxies would form in the spaces in between from matter that was continually being created. The universe existed and would continue to exist forever in more or less the same state as it is today.

For the universe to continue to expand and new matter be created, the steady state model required a modification of general relativity, but the rate of creation needed was very low: about one particle per cubic kilometer per year, which would not conflict with observation. The theory also predicted that the average density of galaxies and similar objects should be constant both in space and time. However, a survey of sources of radio waves outside our galaxy, carried out by Martin Ryle and his group at Cambridge, showed that there were many more faint sources than strong ones. On average, one would expect the faint sources to be the more distant ones. There were thus two possibilities: Either we are in a region of the universe in which strong sources are less frequent than the average; or the density of sources was higher in the past, when the light left the more distant sources on its journey toward us. Neither of these possibilities was compatible with the prediction of the steady state theory that the density of radio sources should be constant in space and time. The final blow to the theory was the discovery in 1964 by Arno Penzias and Robert Wilson of a background of microwave radiation from far beyond our galaxy. This had the characteristic spectrum of radiation emitted by a hot body, though in this case the term *hot* is hardly appropriate, since the temperature was only 2.7 degrees above absolute zero. The universe is a cold, dark place! There was no reasonable mechanism in the steady state theory to generate microwaves with such a spectrum. The theory therefore had to be abandoned.

Another idea that would avoid a big bang singularity was suggested by two Russian scientists, Evgenii Lifshitz and Isaac Khalatnikov, in 1963. They said that a state of infinite density might occur only if the galaxies were moving directly toward or away from each other; only then would they all have met up at a single point in the past. However, the galaxies would also have had some small sideways velocities, and this might have made it possible for there to have been an earlier contracting phase of the universe, in which the galaxies might have come very close together but somehow managed to avoid hitting each other. The universe might then have re-expanded without going through a state of infinite density.

When Lifshitz and Khalatnikov made their suggestion, I was a research student looking for a problem with which to complete my Ph.D. thesis. I was interested in the question of whether there had been a big bang singularity, because that was crucial to an understanding of the origin of the universe. Together with Roger Penrose, I developed a new set of mathematical techniques for dealing with this and similar problems. We showed that if general relativity is correct, any reasonable model of the universe must start with a singularity. This would mean that science could predict that the universe must have had a beginning, but that it could not predict how the universe *should* begin: For that, one would have to appeal to God.

It has been interesting to watch the change in the climate of opinion on singularities. When I was a graduate student, almost no one took them seriously. Now, as a result of the singularity theorems, nearly everyone believes that the universe began with a singularity, at which the laws of physics broke down. However, I now think that although there is a singularity, the laws of physics can still determine how the universe began.

The general theory of relativity is what is called a classical theory. That is, it does not take into account the fact that particles do not have precisely defined positions and velocities but are "smeared out" over a small region by the uncertainty principle of quantum mechanics that does not allow us to measure simultaneously both the position and the velocity. This does not matter in normal situations, because the radius of curvature of space-time is very large compared to the uncertainty in the position of a particle. However, the singularity theorems indicate that space-time will be highly distorted, with a small radius of curvature at the beginning of the present expansion phase of the universe. In this situation, the uncertainty principle will be very important. Thus, general relativity brings about its own downfall by predicting singularities. In order to discuss the beginning of the universe, we need a theory that combines general relativity with quantum mechanics.

That theory is quantum gravity. We do not yet know the exact form the correct theory of quantum gravity will take. The best candidate we have at the moment is the theory of superstrings, but there are still a number of unresolved difficulties. However, certain features can be expected to be present in any viable theory. One is Einstein's idea that the effects of gravity can be represented by a space-time that is curved or distorted—warped—by the matter and energy in it. Objects try to follow the nearest thing to a straight line in this curved space. However, because it is curved their paths appear to be bent, as if by a gravitational field.

Another element that we expect to be present in the ultimate theory is Richard Feynman's proposal that quantum theory can be formulated as a "sum over histories." In its simplest form, the idea is that every particle has every possible path, or history, in space-time. Each path or history has a probability that depends on its shape. For this idea to work, one has to consider histories that take place in imaginary time, rather than in the real time in which we perceive ourselves as living. Imaginary time may sound like something out of science fiction, but it is a well-defined mathematical concept. In a sense it can be thought of as a direction of time that is at right angles to real time. One adds up the probabilities for all the particle histories with certain properties, such as passing through certain points at certain times. One then has to extrapolate the result back to the real space-time in which we live. This is not the most familiar approach to quantum theory, but it gives the same results as other methods.

In the case of quantum gravity, Feynman's idea of a sum over histories would involve summing over different possible histories for the universe: that is, different curved space-times. These would represent the history of the universe and everything in it. One has to specify what class of possible curved spaces should be included in the sum over histories. The choice of this class of spaces determines what state the universe is in. If the class of curved spaces that defines the state of the universe included spaces with singularities, the probabilities of such spaces would not be determined by the theory. Instead, the probabilities would have to be assigned in some arbitrary way. What this means is that science could not predict the probabilities of such singular histories for space-time. Thus, it could not predict how the universe should behave. It is possible, however, that the universe is in a state defined by a sum that includes only nonsingular curved spaces. In this case, the laws of science would determine the universe completely; one would not have to appeal to some agency external to the universe to determine how it began. In a way the proposal that the state of the universe is determined by a sum over only nonsingular histories is like the drunk looking for his key under the lamppost: It may not be where he lost it, but it is the only place where he might find it. Similarly, the universe may not be in the state defined by a sum over nonsingular histories, but it is the only state in which science could predict how the universe should be.

In 1983, Jim Hartle and I proposed that the state of the universe should be given by a sum over a certain class of histories. This class consisted of curved spaces without singularities, which were of finite size but which did not have boundaries or edges. They would be like the surface of the earth but with two more dimensions. The surface of the earth has a finite area, but it doesn't have any singularities, boundaries, or edges. I have tested this by experiment. I went around the world, and I didn't fall off.

The proposal that Hartle and I made can be paraphrased as: The boundary condition of the universe is that it has no boundary. It is only if the universe is in this no-boundary state that the laws of science, on their own, determine the probabilities of each possible history. Thus, it is only in this case that the known laws would determine how the universe should behave. If the universe is in any other state, the class of curved spaces in the sum over histories will include spaces with singularities. In order to determine the probabilities of such singular histories, one would have to invoke some principle other than the known laws of science. This principle would be something external to our universe. We could not deduce it from within our universe. On the other hand, if the universe is in the no-boundary state, we could, in principle, determine completely how the universe should behave, up to the limits of the uncertainty principle.

It would clearly be nice for science if the universe were in the no-boundary state, but how can we tell whether it is? The answer is that the no-boundary proposal makes definite predictions for how the universe should behave. If these predictions were not to agree with observation, we could conclude that the universe is not in the no-boundary state. Thus, the no-boundary proposal is a good scientific theory in the sense defined by the philosopher Karl Popper: It can be disproved or falsified by observation.

If the observations do not agree with the predictions, we will know that there must be singularities in the class of possible histories. However, that is about all we would know. We would not be able to calculate the probabilities of the singular histories; thus, we would not be able to predict how the universe should behave. One might think that this unpredictability wouldn't matter too much if it occurred only at the big bang; after all, that was ten or twenty billion years ago. But if predictability broke down in the very strong gravitational fields in the big bang, it could also break down whenever a star collapsed. This could happen several times a week in our galaxy alone. Our power of prediction would be poor even by the standards of weather forecasts.

Of course, one could say one need not care about the breakdown in predictability that occurred in a distant star. However, in quantum theory, anything that is not actually forbidden can and will happen. Thus, if the class of possible histories includes spaces with singularities, these singularities could occur anywhere, not just at the big bang and in collapsing stars. This would mean that we couldn't predict anything. Conversely, the fact that we are able to predict events is experimental evidence against singularities and for the no-boundary proposal.

So what does the no-boundary proposal predict for the universe? The first point to make is that because all the possible histories for the universe are finite in extent, any quantity that one uses as a measure of time will have a greatest and a least value. Thus, the universe will have a beginning and an end. The beginning in real time will be the big bang singularity. However, the beginning in imaginary time will not be a singularity. Instead, it will be a bit like the North Pole of the earth. If one takes degrees of latitude on the surface of time to be the analogue of time, one could say that the surface of the earth begins at the North Pole. Yet the North Pole is a perfectly ordinary point on the earth. There's nothing special about it, and the same laws hold at the North Pole as at other places on the earth. Similarly, the event that we might choose to label as "the beginning of the universe in imaginary time" would be an ordinary point of space-time, much like any other. The laws of science would hold at the beginning, as elsewhere.

From the analogy with the surface of the earth, one might expect that the end of the universe would be similar to the beginning, just as the North Pole is much like the South Pole. However, the North and South poles correspond to the beginning and end of the history of the universe in imaginary time, not in the real time that we experience. If one extrapolates the results of the sum over histories from imaginary time to real time, one finds that the beginning of the universe in real time can be very different from its end.

Jonathan Halliwell and I have made an approximate calculation of what the no-boundary condition would imply. We treated the universe as a perfectly smooth and uniform background, on which there were small perturbations of density. In real time, the universe would appear to begin its expansion at a very small radius. At first, the expansion would be what is called inflationary: that is, the universe would double in size every tiny fraction of a second, just as prices double every year in certain countries. The world record for economic inflation was probably Germany after the First World War, where the price of a loaf of bread went from under a mark to millions of marks in a few months. But that is nothing compared to the inflation that seems to have occurred in the early universe: an increase in size by a factor of at least a million million million million million times in a tiny fraction of a second. Of course, that was before the present government.

The inflation was a good thing in that it produced a universe that was smooth and uniform on a large scale and was expanding at just the critical rate to avoid recollapse. The inflation was also a good thing in that it produced all the contents of the universe quite literally out of nothing. When the universe was a single point, like the North Pole, it contained nothing. Yet there are now at least ten-to-the-eightieth particles in the part of the universe that we can observe. Where did all these particles come from? The answer is that relativity and quantum mechanics allow matter to be created out of energy in the form of particle/antiparticle pairs. And where did the energy come from to create this matter? The answer is that it was borrowed from the gravitational energy of the universe. The universe has an enormous debt of negative gravitational energy, which exactly balances the positive energy of the matter. During the inflationary period the universe borrowed heavily from its gravitational energy to finance the creation of more matter. The result was a triumph for Keynesian economics: a vigorous and expanding universe, filled with material objects. The debt of gravitational energy will not have to be paid until the end of the universe.

The early universe could not have been completely homogeneous and uniform because that would violate the uncertainty principle of quantum mechanics. Instead, there must have been departures from uniform density. The no-boundary proposal implies that these differences in density would start off in their ground state; that is, they would be as small as possible, consistent with the uncertainty principle. During the inflationary expansion, however, the differences would be amplified. After the period of inflationary expansion was over, one would be left with a universe that was expanding slightly faster in some places than in others. In regions of slower expansion, the gravitational attraction of the matter would slow down the expansion still further. Eventually, the region would stop expanding and would contract to form galaxies and stars. Thus, the no-boundary proposal can account for all the complicated structure that we see around us. However, it does not make just a single prediction for the universe. Instead, it predicts a whole family of possible histories, each with its own probability. There might be a possible history in which the Labour party won the last election in Britain, though maybe the probability is low.

The no-boundary proposal has profound implications for the role of God in the affairs of the universe. It is now generally accepted that the universe evolves according to well-defined laws. These laws may have been ordained by God, but it seems that He does not intervene in the universe to break the laws. Until recently, however, it was thought that these laws did not apply to the beginning of the universe. It would be up to God to wind up the clockwork and set the universe going in any way He wanted. Thus, the present state of the universe would be the result of God's choice of the initial conditions.

The situation would be very different, however, if something like the no-boundary proposal were correct. In that case the laws of physics would hold even at the beginning of the universe, so God would not have had the freedom to choose the initial conditions. Of course, He would still have been free to choose the laws that the universe obeyed. However, this may not have been much of a choice. There may only be a small number of laws, which are self-consistent and which lead to complicated beings like ourselves who can ask the question: What is the nature of God?

And even if there is only one unique set of possible laws, it is only a set of equations. What is it that breathes fire into the equations and makes a universe for them to govern? Is the ultimate unified theory so compelling that it brings about its own existence? Although science may solve the problem of how the universe began, it cannot answer the question: Why does the universe bother to exist? I don't know the answer to that.

Ten

THE QUANTUM MECHANICS OF BLACK HOLES*

T HE FIRST THIRTY years of this century saw the emergence of three theories that radically altered man's view of physics and of reality itself. Physicists are still trying to explore their implications and to fit them together. The three theories are the special theory of relativity (1905), the general theory of relativity (1915), and the theory of quantum mechanics (c. 1926). Albert Einstein was largely responsible for the first, was entirely responsible for the second, and played a major role in the development of the third. Yet Einstein never accepted quantum mechanics because of its element of chance and un-

*An article published in *Scientific American* in January 1977.

certainty. His feelings were summed up in his oft-quoted statement "God does not play dice." Most physicists, however, readily accepted both special relativity and quantum mechanics because they described effects that could be directly observed. General relativity, on the other hand, was largely ignored because it seemed too complicated mathematically, was not testable in the laboratory, and was a purely classical theory that did not seem compatible with quantum mechanics. Thus, general relativity remained in the doldrums for nearly fifty years.

The great extension of astronomical observations that began early in the 1960s brought about a revival of interest in the classical theory of general relativity because it seemed that many of the new phenomena that were being discovered, such as quasars, pulsars, and compact X-ray sources, indicated the existence of very strong gravitational fields—fields that could be described only by general relativity. Quasars are starlike objects that must be many times brighter than entire galaxies if they are as distant as the reddening of their spectra indicates; pulsars are the rapidly blinking remnants of supernova explosions, believed to be ultradense neutron stars; compact X-ray sources, revealed by instruments aboard space vehicles, may also be neutron stars or may be hypothetical objects of still higher density, namely black holes.

One of the problems facing physicists who sought to apply general relativity to these newly discovered or hypothetical objects was to make it compatible with quantum mechanics. Within the past few years there have been developments that give rise to the hope that before too long we shall have a fully consistent quantum theory of gravity, one that will agree with general relativity for macroscopic objects and will, one hopes, be free of the mathematical infinities that have long bedeviled other quantum field theories. These developments have to do with certain recently discovered quantum effects associated with black holes, which provide a remarkable connection between black holes and the laws of thermodynamics.

Let me describe briefly how a black hole might be created. Imagine a star with a mass ten times that of the sun. During most of its lifetime of about a billion years, the star will generate heat at its center by converting oxygen into helium. The energy released will create sufficient pressure to support the star against its own gravity, giving rise to an object with a radius about five times the radius of the sun. The escape velocity from the surface of such a star would be about a thousand kilometers per second. That is to say, an object fired vertically upward from the surface of the star with a velocity of less than a thousand kilometers per second would be dragged back by the gravitational field of the star and would return to the surface, whereas an object with a velocity greater than that would escape to infinity.

When the star had exhausted its nuclear fuel, there would be nothing to maintain the outward pressure, and the star would begin to collapse because of its own gravity. As the star shrank, the gravitational field at the surface would become stronger and the escape velocity would increase. By the time the radius had got down to thirty kilometers, the escape velocity would have increased to 300,000 kilometers per second, the velocity of light. After that time any light emitted from the star would not be able to escape to infinity but would be dragged back by the gravitational field. According to the special theory of relativity, nothing can travel faster than light, so that if light cannot escape, nothing else can either.

The result would be a black hole: a region of space-time from which it is not possible to escape to infinity. The boundary of the black hole is called the event horizon. It corresponds to a wave front of light from the star that just fails to escape to infinity but remains hovering at the Schwarzschild radius:

$2\ GM/\sqrt{c}$, where G is Newton's constant of gravity, M is the mass of the star, and c is the velocity of light. For a star of about ten solar masses, the Schwarzschild radius is about thirty kilometers.

There is now fairly good observational evidence to suggest that black holes of about this size exist in double-star systems such as the X-ray source known as Cygnus X-I. There might also be quite a number of very much smaller black holes scattered around the universe, formed not by the collapse of stars but by the collapse of highly compressed regions in the hot, dense medium that is believed to have existed shortly after the big bang in which the universe originated. Such "primordial" black holes are of greatest interest for the quantum effects I shall describe here. A black hole weighing a billion tons (about the mass of a mountain) would have a radius of about 10^{-13} centimeter (the size of a neutron or a proton). It could be in orbit either around the sun or around the center of the galaxy.

The first hint that there might be a connection between black holes and thermodynamics came with the mathematical discovery in 1970 that the surface area of the event horizon, the boundary of a black hole, has the property that it always increases when additional matter or radiation falls into the black hole. Moreover, if two black holes collide and merge to form a single black hole, the area of the event horizon around the resulting black hole is greater than the sum of the areas of the event horizons around the original black holes. These properties suggest that there is a resemblance between the area of the event horizon of a black hole and the concept of entropy in thermodynamics. Entropy can be regarded as a measure of the disorder of a system or, equivalently, as a lack of knowledge of its precise state. The famous second law of thermodynamics says that entropy always increases with time.

The analogy between the properties of black holes and the

laws of thermodynamics has been extended by James M. Bardeen of the University of Washington, Brandon Carter, who is now at the Meudon Observatory, and me. The first law of thermodynamics says that a small change in the entropy of a system is accompanied by a proportional change in the energy of the system. The factor of proportionality is called the temperature of the system. Bardeen, Carter, and I found a similar law relating to the change in mass of a black hole to a change in the area of the event horizon. Here the factor of proportionality involves a quantity called the surface gravity, which is a measure of the strength of the gravitational field at the event horizon. If one accepts that the area of the event horizon is analogous to entropy, then it would seem that the surface gravity is analogous to temperature. The resemblance is strengthened by the fact that the surface gravity turns out to be the same at all points on the event horizon, just as the temperature is the same everywhere in a body at thermal equilibrium.

Although there is clearly a similarity between entropy and the area of the event horizon, it was not obvious to us how the area could be identified as the entropy of a black hole. What would be meant by the entropy of a black hole? The crucial suggestion was made in 1972 by Jacob D. Bekenstein, who was then a graduate student at Princeton University and is now at the University of the Negev in Israel. It goes like this. When a black hole is created by gravitational collapse, it rapidly settles down to a stationary state that is characterized by only three parameters: the mass, the angular momentum, and the electric charge. Apart from these three properties the black hole preserves no other details of the object that collapsed. This conclusion, known as the theorem "A black hole has no hair," was proved by the combined work of Carter, Werner Israel of the University of Alberta, David C. Robinson of King's College, London, and me.

The no-hair theorem implies that a large amount of information is lost in a gravitational collapse. For example, the final black-hole state is independent of whether the body that collapsed was composed of matter or antimatter, and whether it was spherical or highly irregular in shape. In other words, a black hole of a given mass, angular momentum, and electric charge could have been formed by the collapse of any one of a large number of different configurations of matter. Indeed, if quantum effects are neglected, the number of configurations would be infinite, since the black hole could have been formed by the collapse of a cloud of an indefinitely large number of particles of indefinitely low mass.

The uncertainty principle of quantum mechanics implies, however, that a particle of mass m behaves like a wave of wavelength h/mc, where h is Planck's constant (the small number 6.62×10^{-27} erg-second) and c is the velocity of light. In order for a cloud of particles to be able to collapse to form a black hole, it would seem necessary for this wavelength to be smaller than the size of the black hole that would be formed. It therefore appears that the number of configurations that could form a black hole of a given mass, angular momentum, and electric charge, although very large, may be finite. Bekenstein suggested that one could interpret the logarithm of this number as the entropy of a black hole. The logarithm of the number would be a measure of the amount of information that was irretrievably lost during the collapse through the event horizon when a black hole was created.

The apparently fatal flaw in Bekenstein's suggestion was that if a black hole has a finite entropy that is proportional to the area of its event horizon, it also ought to have a finite temperature, which would be proportional to its surface gravity. This would imply that a black hole could be in equilibrium with thermal radiation at some temperature other than zero. Yet according to classical concepts no such equilibrium is possible, since the black hole would absorb any thermal radiation that fell on it but by definition would not be able to emit anything in return.

This paradox remained until early 1974, when I was investigating what the behavior of matter in the vicinity of a black hole would be according to quantum mechanics. To my great surprise, I found that the black hole seemed to emit particles at a steady rate. Like everyone else at that time, I accepted the dictum that a black hole could not emit anything. I therefore put quite a lot of effort into trying to get rid of this embarrassing effect. It refused to go away, so that in the end I had to accept it. What finally convinced me that it was a real physical process was that the outgoing particles have a spectrum that is precisely thermal; the black hole creates and emits particles just as if it were an ordinary hot body with a temperature that is proportional to the surface gravity and inversely proportional to the mass. This made Bekenstein's suggestion that a black hole had a finite entropy fully consistent, since it implied that a black hole could be in thermal equilibrium at some finite temperature other than zero.

Since that time, the mathematical evidence that black holes can emit thermally has been confirmed by a number of other people with various different approaches. One way to understand the emission is as follows. Quantum mechanics implies that the whole of space is filled with pairs of "virtual" particles and antiparticles that are constantly materializing in pairs, separating, and then coming together again and annihilating each other. These particles are called virtual because, unlike "real" particles, they cannot be observed directly with a particle detector. Their indirect effects can nonetheless be measured, and their existence has been confirmed by a small shift (the "Lamb shift") they produce in the spectrum of light from excited hydrogen atoms. Now, in the presence of a black hole one member of a pair of virtual particles may fall into the hole, leaving the other member without a partner with which to annihilate. The forsaken particle or antiparticle may fall into the black hole after its partner, but it may also escape to infinity, where it appears to be radiation emitted by the black hole.

Another way of looking at the process is to regard the member of the pair of particles that falls into the black hole—the antiparticle, say—as being really a particle that is traveling backward in time. Thus, the antiparticle falling into the black hole can be regarded as a particle coming out of the black hole but traveling backward in time. When the particle reaches the point at which the particle-antiparticle pair originally materialized, it is scattered by the gravitational field so that it travels forward in time.

Quantum mechanics therefore allows a particle to escape from inside a black hole, something that is not allowed in classical mechanics. There are, however, many other situations in atomic and nuclear physics where there is some kind of barrier that particles should not be able to penetrate on classical principles but that they are able to tunnel through on quantum-mechanical principles.

The thickness of the barrier around a black hole is proportional to the size of the black hole. This means that very few particles can escape from a black hole as large as the one hypothesized to exist in Cygnus X-I, but that particles can leak very rapidly out of smaller black holes. Detailed calculations show that the emitted particles have a thermal spectrum corresponding to a temperature that increases rapidly as the mass of the black hole decreases. For a black hole with a mass of the sun, the temperature is only about a ten-millionth of a degree above absolute zero. The thermal radiation leaving a black hole with that temperature would be completely swamped by the general background of radiation in the universe. On the other hand, a black hole with a mass of only a billion tons—that is, a primordial black hole, roughly the size of a proton—would have a temperature of some 120 billion degrees Kelvin, which corresponds to an energy of some ten million electron volts. At such a temperature a black hole would be able to create electron-positron pairs and particles of zero mass, such as photons, neutrinos, and gravitons (the presumed carriers of gravitational energy). A primordial black hole would release energy at the rate of 6,000 megawatts, equivalent to the output of six large nuclear power plants.

As a black hole emits particles, its mass and size steadily decrease. This makes it easier for more particles to tunnel out, and so the emission will continue at an ever-increasing rate until eventually the black hole radiates itself out of existence. In the long run, every black hole in the universe will evaporate in this way. For large black holes, however, the time it will take is very long indeed; a black hole with the mass of the sun will last for about 10^{66} years. On the other hand, a primordial black hole should have almost completely evaporated in the ten billion years that have elapsed since the big bang, the beginning of the universe as we know it. Such black holes should now be emitting hard gamma rays with an energy of about 100 million electron volts.

Calculations made by Don N. Page, then of the California Institute of Technology, and me, based on measurements of the cosmic background of gamma radiation made by the satellite SAS-2, show that the average density of primordial black holes in the universe must be less than about two hundred per cubic light-year. The local density in our galaxy could be a million times higher than this figure if primordial black holes were concentrated in the "halo" of galaxies—the thin cloud of rapidly moving stars in which each galaxy is embedded—rather than being uniformly distributed throughout the universe. This would imply that the primordial black hole closest to the earth is probably at least as far away as the planet Pluto.

The final stage of the evaporation of a black hole would proceed so rapidly that it would end in a tremendous explosion. How powerful this explosion would be would depend on how many different species of elementary particles there are. If, as is now widely believed, all particles are made up of perhaps six different varieties of quarks, the final explosion would have an energy equivalent to about ten million one-megaton hydrogen bombs. On the other hand, an alternative theory put forward by R. Hagedorn of CERN, the European Organization for Nuclear Research in Geneva, argues that there is an infinite number of elementary particles of higher and higher mass. As a black hole got smaller and hotter, it would emit a larger and larger number of different species of particles and would produce an explosion perhaps 100,000 times more powerful than the one calculated on the quark hypothesis. Hence the observation of a black-hole explosion would provide very important information on elementary particle physics, information that might not be available any other way.

A black-hole explosion would produce a massive outpouring of high-energy gamma rays. Although they might be observed by gamma-ray detectors on satellites or balloons, it would be difficult to fly a detector large enough to have a reasonable chance of intercepting a significant number of gamma-ray photons from one explosion. One possibility would be to employ a space shuttle to build a large gamma-ray detector in orbit. An easier and much cheaper alternative would be to let the earth's upper atmosphere serve as a detector. A high-energy gamma ray plunging into the atmosphere will create a shower of electron-positron pairs, which initially will be traveling through the atmosphere faster than light can. (Light is slowed down by interactions with the air molecules.) Thus the electrons and positrons will set up a kind of sonic boom, or shock wave, in the electromagnetic field. Such a shock wave, called Cerenkov radiation, could be detected from the ground as a flash of visible light.

A preliminary experiment by Neil A. Porter and Trevor C. Weekes of University College, Dublin, indicates that if black holes explode the way Hagedorn's theory predicts, there are fewer than two black-hole explosions per cubic light-year per century in our region of the galaxy. This would imply that the density of primordial black holes is less than 100 million per cubic light-year. It should be possible to greatly increase the sensitivity of such observations. Even if they do not yield any positive evidence of primordial black holes, they will be very valuable. By placing a low upper limit on the density of such black holes, the observations will indicate that the early universe must have been very smooth and nonturbulent.

The big bang resembles a black-hole explosion but on a vastly larger scale. One therefore hopes that an understanding of how black holes create particles will lead to a similar understanding of how the big bang created everything in the universe. In a black hole, matter collapses and is lost forever, but new matter is created in its place. It may therefore be that there was an earlier phase of the universe in which matter collapsed, to be re-created in the big bang.

If the matter that collapses to form a black hole has a net electric charge, the resulting black hole will carry the same charge. This means that the black hole will tend to attract those members of the virtual particle-antiparticle pairs that have the opposite charge and repel those that have a like charge. The black hole will therefore preferentially emit particles with charge of the same sign as itself and so will rapidly lose its charge. Similarly, if the collapsing matter has a net angular momentum, the resulting black hole will be rotating and will preferentially emit particles that carry away its angular momentum. The reason a black hole "remembers" the electric charge, angular momentum, and mass of the matter that collapsed and "forgets" everything else is that these three quantities are coupled to long-range fields: in the case of charge the electromagnetic field, and in the case of angular momentum and mass the gravitational field.

Experiments by Robert H. Dicke of Princeton University and Vladimir Braginsky of Moscow State University have indicated that there is no long-range field associated with the quantum property designated baryon number. (Baryons are the class of particles including the proton and the neutron.) Hence, a black hole formed out of the collapse of a collection of baryons would forget its baryon number and radiate equal quantities of baryons and antibaryons. Therefore, when the black hole disappeared, it would violate one of the most cherished laws of particle physics, the law of baryon conservation.

Although Bekenstein's hypothesis that black holes have a finite entropy requires for its consistency that black holes should radiate thermally, at first it seems a complete miracle that the detailed quantum-mechanical calculations of particle creation should give rise to emission with a thermal spectrum. The explanation is that the emitted particles tunnel out of the black hole from a region of which an external observer has no knowledge other than its mass, angular momentum, and electric charge. This means that all combinations or configurations of emitted particles that have the same energy, angular momentum, and electric charge are equally probable. Indeed, it is possible that the black hole could emit a television set or the works of Proust in ten leatherbound volumes, but the number of configurations of particles that correspond to these exotic possibilities is vanishingly small. By far the largest number of configurations correspond to emission with a spectrum that is nearly thermal.

The emission from black holes has an added degree of uncertainty, or unpredictability, over and above that normally associated with quantum mechanics. In classical mechanics one can predict the results of measuring both the position and the velocity of a particle. In quantum mechanics the uncertainty principle says that only one of these measurements can be predicted; the observer can predict the result of measuring either the position or the velocity but not both. Alternatively, he can predict the result of measuring one combination of position and velocity. Thus, the observer's ability to make definite predictions is in effect cut in half. With black holes the situation is even worse. Since the particles emitted by a black hole come from a region of which the observer has very limited knowledge, he cannot definitely predict the position or the velocity of a particle or any combination of the two; all he can predict is the probabilities that certain particles will be emitted. It therefore seems that Einstein was doubly wrong when he said, "God does not play dice." Consideration of particle emission from black holes would seem to suggest that God not only plays dice but also sometimes throws them where they cannot be seen.

Eleven

BLACK HOLES
AND
BABY UNIVERSES*

F ALLING INTO A black hole has become one of the hor-
rors of science fiction. In fact, black holes can now be
said to be really matters of science fact rather than science fiction. As I shall describe, there are good reasons for predicting
that black holes should exist, and the observational evidence
points strongly to the presence of a number of black holes in
our own galaxy and more in other galaxies.

Of course, where the science fiction writers really go to
town is on what happens if you do fall in a black hole. A common suggestion is that if the black hole is rotating, you can fall

*Hitchcock lecture, given at the University of California, Berkeley, in April 1988.

through a little hole in space-time and out into another region of the universe. This obviously raises great possibilities for space travel. Indeed, we will need something like this if travel to other stars, let alone to other galaxies, is to be a practical proposition in the future. Otherwise, the fact that nothing can travel faster than light means that the round trip to the nearest star would take at least eight years. So much for a weekend break on Alpha Centauri! On the other hand, if one could pass through a black hole, one might reemerge anywhere in the universe. Quite how to choose your destination is not clear: You might set out for a holiday in Virgo and end up in the Crab Nebula.

I'm sorry to disappoint prospective galactic tourists, but this scenario doesn't work: If you jump into a black hole, you will get torn apart and crushed out of existence. However, there is a sense in which the particles that make up your body will carry on into another universe. I don't know if it would be much consolation to someone being made into spaghetti in a black hole to know that his particles might survive.

Despite the slightly flippant tone I have adopted, this essay is based on hard science. Most of what I say here is now agreed upon by other scientists working in this field, though this acceptance has come only fairly recently. The last part of the essay, however, is based on very recent work on which there is, as yet, no general consensus. But this work is arousing great interest and excitement.

Although the concept of what we now call a black hole goes back more than two hundred years, the name *black hole* was introduced only in 1967 by the American physicist John Wheeler. It was a stroke of genius: The name ensured that black holes entered the mythology of science fiction. It also stimulated scientific research by providing a definite name for something that previously had not had a satisfactory title.

The importance in science of a good name should not be underestimated.

As far as I know, the first person to discuss black holes was a Cambridge man called John Michell, who wrote a paper about them in 1783. His idea was this: Suppose you fire a cannonball vertically upward from the surface of the earth. As it goes up, it will be slowed down by the effect of gravity. Eventually, it will stop going up and will fall back to earth. If it started with more than a certain critical speed, however, it would never stop rising and fall back but would continue to move away. This critical speed is called the escape velocity. It is about seven miles a second for the earth, and about one hundred miles a second for the sun. Both of these velocities are greater than the speed of a real cannonball, but they are much smaller than the velocity of light, which is 186,000 miles a second. This means that gravity doesn't have much effect on light; light can escape without difficulty from the earth or the sun. However, Michell reasoned that it would be possible to have a star that was sufficiently massive and sufficiently small in size that its escape velocity would be greater than the velocity of light. We would not be able to see such a star because light from its surface would not reach us; it would be dragged back by the star's gravitational field. However, we might be able to detect the presence of the star by the effect that its gravitational field would have on nearby matter.

It is not really consistent to treat light like cannonballs. According to an experiment carried out in 1897, light always travels at the same constant velocity. How then can gravity slow down light? A consistent theory of how gravity affects light did not come until 1915, when Einstein formulated the general theory of relativity. Even so, the implications of this theory for old stars and other massive bodies were not generally realized until the 1960s.

According to general relativity, space and time together can be regarded as forming a four-dimensional space called space-time. This space is not flat; it is distorted, or curved, by the matter and energy in it. We observe this curvature in the bending of the light or radio waves that travel near the sun on their way to us. In the case of light passing near the sun, the bending is very small. However, if the sun were to shrink until it was only a few miles across, the bending would be so great that light leaving the sun would not get away but would be dragged back by the sun's gravitational field. According to the theory of relativity, nothing can travel faster than the speed of light, so there would be a region from which it would be impossible for anything to escape. This region is called a black hole. Its boundary is called the event horizon. It is formed by the light that just fails to get away from the black hole but stays hovering on the edge.

It might sound ridiculous to suggest that the sun could shrink to being only a few miles across. One might think that matter could not be compressed that far. But it turns out that it can.

The sun is the size it is because it is so hot. It is burning hydrogen into helium, like a controlled H-bomb. The heat released in this process generates a pressure that enables the sun to resist the attraction of its own gravity, which is trying to make it smaller.

Eventually, however, the sun will run out of nuclear fuel. This will not happen for about another five billion years, so there's no great rush to book your flight to another star. However, stars more massive than the sun will burn up their fuel much more rapidly. When they finish their fuel, they will start to lose heat and contract. If they are less than about twice the mass of the sun, they will eventually stop contracting and will settle down to a stable state. One such state is called a white dwarf. These have radii of a few thousand miles and densities of hundreds of tons per cubic inch. Another such state is a neutron star. These have a radius of about ten miles and densities of millions of tons per cubic inch.

We observe large numbers of white dwarfs in our immediate neighborhood in the galaxy. Neutron stars, however, were not observed until 1967, when Jocelyn Bell and Antony Hewish at Cambridge discovered objects called pulsars that were emitting regular pulses of radio waves. At first, they wondered whether they had made contact with an alien civilization; indeed, I remember that the seminar room in which they announced their discovery was decorated with figures of "little green men." In the end, however, they and everyone else came to the less romantic conclusion that these objects were rotating neutron stars. This was bad news for writers of space Westerns but good news for the small number of us who believed in black holes at that time. If stars could shrink to as small as ten or twenty miles across to become neutron stars, one might expect that other stars could shrink even further to become black holes.

A star with a mass more than about twice that of the sun cannot settle down as a white dwarf or neutron star. In some cases, the star may explode and throw off enough matter to bring its mass below the limit. But this won't happen in all cases. Some stars will become so small that their gravitational fields will bend light to that point that it comes back toward the star. No further light, or anything else, will be able to escape. The stars will have become black holes.

The laws of physics are time-symmetric. So if there are objects called black holes into which things can fall but not get out, there ought to be other objects that things can come out of but not fall into. One could call these white holes. One might speculate that one could jump into a black hole in one place and come out of a white hole in another. This would be the ideal method of long-distance space travel mentioned earlier. All you would need would be to find a nearby black hole.

At first, this form of space travel seemed possible. There are solutions of Einstein's general theory of relativity in which it is possible to fall into a black hole and come out of a white hole. Later work, however, shows that these solutions are all very unstable: the slightest disturbance, such as the presence of a spaceship, would destroy the "wormhole," or passage, leading from the black hole to the white hole. The spaceship would be torn apart by infinitely strong forces. It would be like going over Niagara in a barrel.

After that, it seemed hopeless. Black holes might be useful for getting rid of garbage or even some of one's friends. But they were "a country from which no traveler returns." Everything I have been saying so far, however, has been based on calculations using Einstein's general theory of relativity. This theory is in excellent agreement with all the observations we have made. But we know it cannot be quite right because it doesn't incorporate the uncertainty principle of quantum mechanics. The uncertainty principle says that particles cannot have both a well-defined position and a well-defined velocity. The more precisely you measure the position of a particle, the less precisely you can measure its velocity, and vice versa.

In 1973 I started investigating what difference the uncertainty principle would make to black holes. To my great surprise and that of everyone else, I found that it meant that black holes are not completely black. They would be sending out radiation and particles at a steady rate. My results were received with general disbelief when I announced them at a conference near Oxford. The chairman of the session said they were nonsense, and he wrote a paper saying so. However, when other people repeated my calculation, they found the same effect. So in the end, even the chairman agreed I was right.

How can radiation escape from the gravitational field of a black hole? There are a number of ways one can understand how. Although they seem very different, they are really all equivalent. One way is to realize that the uncertainty principle allows particles to travel faster than light for a short distance. This enables particles and radiation to get out through the event horizon and escape from the black hole. Thus, it is possible for things to get out of a black hole. What comes out of a black hole, however, will be different from what fell in. Only the energy will be the same.

As a black hole gives off particles and radiation, it will lose mass. This will cause the black hole to get smaller and to send out particles more rapidly. Eventually, it will get down to zero mass and will disappear completely. What will happen then to the objects, including possible spaceships, that have fallen into the black hole? According to some recent work of mine, the answer is that they will go off into a little baby universe of their own. A small, self-contained universe branches off from our region of the universe. This baby universe may join on again to our region of space-time. If it does, it would appear to us to be another black hole that formed and then evaporated. Particles that fell into one black hole would appear as particles emitted by the other black hole, and vice versa.

This sounds like just what is required to allow space travel through black holes. You just steer your spaceship into a suitable black hole. It had better be a pretty big one, though, or the gravitational forces will tear you into spaghetti before you get inside. You would then hope to reappear out of some other hole, though you wouldn't be able to choose where.

However, there's a snag in this intergalactic transportation scheme. The baby universes that take the particles that fell into the hole occur in what is called imaginary time. In real time, an astronaut who fell into a black hole would come to a sticky end. He would be torn apart by the difference between the gravitational force on his head and his feet. Even the particles that made up his body would not survive. Their histories, in real time, would come to an end at a singularity. But the histories of the particles in imaginary time would continue. They would pass into the baby universe and would reemerge as the particles emitted by another black hole. Thus, in a sense, the astronaut would be transported to another region of the universe. However, the particles that emerged would not look much like the astronaut. Nor might it be much consolation to him, as he ran into the singularity in real time, to know that his particles will survive in imaginary time. The motto for anyone who falls into a black hole must be: "Think imaginary."

What determines where the particles reemerge? The number of particles in the baby universe will be equal to the number of particles that have fallen into the black hole, plus the number of particles that the black hole emits during its evaporation. This means that the particles that fall into one black hole will come out of another hole of about the same mass. Thus, one might try to select where the particles would come out by creating a black hole of the same mass as that into which the particles went down. However, the black hole would be equally likely to give off any other set of particles with the same total energy. Even if the black hole did emit the right kinds of particles, one could not tell if they were actually the same particles that had gone down the other hole. Particles do not carry identity cards; all particles of a given kind look alike.

What all this means is that going through a black hole is unlikely to prove a popular and reliable method of space travel. First of all, you would have to get there by traveling in imaginary time and not care that your history in real time came to a sticky end. Second, you couldn't really choose your destination. It would be like traveling on some airlines I could name.

Although baby universes may not be of much use for space travel, they have important implications for our attempt to find a complete unified theory that will describe everything in the universe. Our present theories contain a number of quantities, like the size of the electric charge on a particle. The values of these quantities cannot be predicted by our theories. Instead, they have to be chosen to agree with observations. Most scientists believe, however, that there is some underlying unified theory that will predict the values of all these quantities.

There may well be such an underlying theory. The strongest candidate at the moment is called the heterotic superstring. The idea is that space-time is filled with little loops, like pieces of string. What we think of as elementary particles are really these little loops vibrating in different ways. This theory does not contain any numbers whose values can be adjusted. One would therefore expect that this unified theory should be able to predict all the values of quantities, like the electric charge on a particle, that are left undetermined by our present theories. Even though we have not yet been able to predict any of these quantities from superstring theory, many people believe that we will be able to do so eventually.

However, if this picture of baby universes is correct, our ability to predict these quantities will be reduced. This is because we cannot observe how many baby universes exist out there, waiting to join onto our region of the universe. There can be baby universes that contain only a few particles. These baby universes are so small that one would not notice them joining on or branching off. By joining on, however, they will alter the apparent values of quantities, such as the electric charge on a particle. Thus, we will not be able to predict what the apparent values of these quantities will be because we don't know how many baby universes are waiting out there. There could be a population explosion of baby universes. Unlike the human case, however, there seem to be no limiting factors such as food supply or standing room. Baby universes exist in a realm of their own. It is a bit like asking how many angels can dance on the head of a pin.

For most quantities, baby universes seem to introduce a definite, although fairly small, amount of uncertainty in the predicted values. However, they may provide an explanation of the observed value of one very important quantity: the so-called cosmological constant. This is a term in the equations of general relativity that gives space-time an inbuilt tendency to expand or contract. Einstein originally proposed a very small cosmological constant in the hope of balancing the tendency of matter to make the universe contract. That motivation disappeared when it was discovered that the universe is expanding. But it was not so easy to get rid of the cosmological constant. One might expect the fluctuations that are implied by quantum mechanics to give a cosmological constant that is very large. Yet we can observe how the expansion of the universe is varying with time and thus determine that the cosmological constant is very small. Up to now, there has been no good explanation for why the observed value should be so small. However, baby universes branching off and joining on will affect the apparent value of the cosmological constant. Because we don't know how many baby universes there are, there will be different possible values for the apparent cosmological constant. A nearly zero value, however, will be by far the most probable. This is fortunate because it is only if the value of the cosmological constant is very small that the universe would be suitable for beings like us.

To sum up: It seems that particles can fall into black holes that then evaporate and disappear from our region of the universe. The particles go off into baby universes that branch off from our universe. These baby universes can then join back on somewhere else. They may not be much good for space travel, but their presence means that we will be able to predict less than we expected, even if we do find a complete unified theory. On the other hand, we now may be able to provide explanations for the measured values of some quantities like the cosmological constant. In the last few years, a lot of people have begun working on baby universes. I don't think anyone will make a fortune by patenting them as a method of space travel, but they have become a very exciting area of research.

Twelve

IS EVERYTHING DETERMINED?*

I N THE PLAY *Julius Caesar*, Cassius tells Brutus, "Men at some times are masters of their fate." But are we really masters of our fate? Or is everything we do determined and preordained? The argument for preordination used to be that God was omnipotent and outside time, so God would know what was going to happen. But how then could we have any free will? And if we don't have free will, how can we be responsible for our actions? It can hardly be one's fault if one has been preordained to rob a bank. So why should one be punished for it?

*A lecture given at the Sigma Club seminar at the University of Cambridge, April 1990.

In recent times, the argument for determinism has been based on science. It seems that there are well-defined laws that govern how the universe and everything in it develops in time. Although we have not yet found the exact form of all these laws, we already know enough to determine what happens in all but the most extreme situations. Whether we will find the remaining laws in the fairly near future is a matter of opinion. I'm an optimist: I think there's a fifty-fifty chance that we will find them in the next twenty years. But even if we don't, it won't really make any difference to the argument. The important point is that there should exist a set of laws that completely determines the evolution of the universe from its initial state. These laws may have been ordained by God. But it seems that He (or She) does not intervene in the universe to break the laws.

The initial configuration of the universe may have been chosen by God, or it may itself have been determined by the laws of science. In either case, it would seem that everything in the universe would then be determined by evolution according to the laws of science, so it is difficult to see how we can be masters of our fate.

The idea that there is some grand unified theory that determines everything in the universe raises many difficulties. First of all, the grand unified theory is presumably compact and elegant in mathematical terms. There ought to be something special and simple about the theory of everything. Yet how can a certain number of equations account for the complexity and trivial detail that we see around us? Can one really believe that the grand unified theory has determined that Sinead O'Connor will be the top of the hit parade this week, or that Madonna will be on the cover of *Cosmopolitan*?

A second problem with the idea that everything is determined by a grand unified theory is that anything we say is also determined by the theory. But why should it be determined to be correct? Isn't it more likely to be wrong, because there are many possible incorrect statements for every true one? Each week, my mail contains a number of theories that people have sent me. They are all different, and most are mutually inconsistent. Yet presumably the grand unified theory has determined that the authors think they were correct. So why should anything I say have any greater validity? Aren't I equally determined by the grand unified theory?

A third problem with the idea that everything is determined is that we feel that we have free will—that we have the freedom to choose whether to do something. But if everything is determined by the laws of science, then free will must be an illusion, and if we don't have free will, what is the basis for our responsibility for our actions? We don't punish people for crimes if they are insane, because we have decided that they can't help it. But if we are all determined by a grand unified theory, none of us can help what we do, so why should anyone be held responsible for what they do?

These problems of determinism have been discussed over the centuries. The discussion was somewhat academic, however, as we were far from a complete knowledge of the laws of science, and we didn't know how the initial state of the universe was determined. The problems are more urgent now because there is the possibility that we may find a complete unified theory in as little as twenty years. And we realize that the initial state may itself have been determined by the laws of science. What follows is my personal attempt to come to terms with these problems. I don't claim any great originality or depth, but it is the best I can do at the moment.

To start with the first problem: How can a relatively simple and compact theory give rise to a universe that is as complex as the one we observe, with all its trivial and unimportant details? The key to this is the uncertainty principle of quantum mechanics, which states that one cannot measure both the position and speed of a particle to great accuracy; the more accurately you measure the position, the less accurately you can measure the speed, and vice versa. This uncertainty is not so important at the present time, when things are far apart, so that a small uncertainty in position does not make much difference. But in the very early universe, everything was very close together, so there was quite a lot of uncertainty, and there were a number of possible states for the universe. These different possible early states would have evolved into a whole family of different histories for the universe. Most of these histories would be similar in their large-scale features. They would correspond to a universe that was uniform and smooth, and that was expanding. However, they would differ on details like the distribution of stars and, even more, on what was on the covers of their magazines. (That is, if those histories contained magazines.) Thus the complexity of the universe around us and its details arose from the uncertainty principle in the early stages. This gives a whole family of possible histories for the universe. There would be a history in which the Nazis won the Second World War, though the probability is low. But we just happen to live in a history in which the Allies won the war and Madonna was on the cover of *Cosmopolitan*.

I now turn to the second problem: If what we do is determined by some grand unified theory, why should the theory determine that we draw the right conclusions about the universe rather than the wrong ones? Why should anything we say have any validity? My answer to this is based on Darwin's idea of natural selection. I take it that some very primitive form of life arose spontaneously on earth from chance combinations of atoms. This early form of life was probably a large molecule. But it was probably not DNA, since the chances of forming a whole DNA molecule by random combinations are small.

The early form of life would have reproduced itself. The quantum uncertainty principle and the random thermal motions of the atoms would mean that there were a certain number of errors in the reproduction. Most of these errors would have been fatal to the survival of the organism or its ability to reproduce. Such errors would not be passed on to future generations but would die out. A very few errors would be beneficial, by pure chance. The organisms with these errors would be more likely to survive and reproduce. Thus they would tend to replace the original, unimproved organisms.

The development of the double helix structure of DNA may have been one such improvement in the early stages. This was probably such an advance that it completely replaced any earlier form of life, whatever that may have been. As evolution progressed, it would have led to the development of the central nervous system. Creatures that correctly recognized the implications of data gathered by their sense organs and took appropriate action would be more likely to survive and reproduce. The human race has carried this to another stage. We are very similar to higher apes, both in our bodies and in our DNA; but a slight variation in our DNA has enabled us to develop language. This has meant that we can hand down information and accumulated experience from generation to generation, in spoken and eventually in written form. Previously, the results of experience could be handed down only by the slow process of it being encoded into DNA through random errors in reproduction. The effect has been a dramatic speed-up of evolution.

It took more than three billion years to evolve up to the human race. But in the course of the last ten thousand years, we have developed written language. This has enabled us to progress from cave dwellers to the point where we can ask about the ultimate theory of the universe.

There has been no significant biological evolution, or change in human DNA, in the last ten thousand years. Thus, our intelligence, our ability to draw the correct conclusions from the information provided by our sense organs, must date back to our cave dweller days or earlier. It would have been selected for on the basis of our ability to kill certain animals for food and to avoid being killed by other animals. It is remarkable that mental qualities that were selected for these purposes should have stood us in such good stead in the very different circumstances of the present day. There is probably not much survival advantage to be gained from discovering a grand unified theory or answering questions about determinism. Nevertheless, the intelligence that we have developed for other reasons may well ensure that we find the right answers to these questions.

I now turn to the third problem, the questions of free will and responsibility for our actions. We feel subjectively that we have the ability to choose who we are and what we do. But this may just be an illusion. Some people think they are Jesus Christ or Napoleon, but they can't all be right. What we need is an objective test that we can apply from the outside to distinguish whether an organism has free will. For example, suppose we were visited by a "little green person" from another star. How could we decide whether it had free will or was just a robot, programmed to respond as if it were like us?

The ultimate objective test of free will would seem to

be: Can one predict the behavior of the organism? If one can, then it clearly doesn't have free will but is predetermined. On the other hand, if one cannot predict the behavior, one could take that as an operational definition that the organism has free will.

One might object to this definition of free will on the grounds that once we find a complete unified theory we will be able to predict what people will do. The human brain, however, is also subject to the uncertainty principle. Thus, there is an element of the randomness associated with quantum mechanics in human behavior. But the energies involved in the brain are low, so quantum mechanical uncertainty is only a small effect. The real reason why we cannot predict human behavior is that it is just too difficult. We already know the basic physical laws that govern the activity of the brain, and they are comparatively simple. But it is just too hard to solve the equations when there are more than a few particles involved. Even in the simpler Newtonian theory of gravity, one can solve the equations exactly only in the case of two particles. For three or more particles one has to resort to approximations, and the difficulty increases rapidly with the number of particles. The human brain contains about 10^{26} or a hundred million billion billion particles. This is far too many for us ever to be able to solve the equations and predict how the brain would behave, given its initial state and the nerve data coming into it. In fact, of course, we cannot even measure what the initial state was, because to do so we would have to take the brain apart. Even if we were prepared to do that, there would just be too many particles to record. Also, the brain is probably very sensitive to the initial state—a small change in the initial state can make a very large difference to subsequent behavior. So although we know the fundamental equations that govern the brain, we are quite unable to use them to predict human behavior.

This situation arises in science whenever we deal with the macroscopic system, because the number of particles is always too large for there to be any chance of solving the fundamental equations. What we do instead is use effective theories. These are approximations in which the very large number of particles are replaced by a few quantities. An example is fluid mechanics. A liquid such as water is made up of billions of billions of molecules that themselves are made up of electrons, protons, and neutrons. Yet it is a good approximation to treat the liquid as a continuous medium, characterized just by velocity, density, and temperature. The predictions of the effective theory of fluid mechanics are not exact—one only has to listen to the weather forecast to realize that—but they are good enough for the design of ships or oil pipelines.

I want to suggest that the concepts of free will and moral responsibility for our actions are really an effective theory in the sense of fluid mechanics. It may be that everything we do is determined by some grand unified theory. If that theory has determined that we shall die by hanging, then we shall not drown. But you would have to be awfully sure that you were destined for the gallows to put to sea in a small boat during a storm. I have noticed that even people who claim that everything is predestined and that we can do nothing to change it look before they cross the road. Maybe it's just that those who don't look don't survive to tell the tale.

One cannot base one's conduct on the idea that everything is determined, because one does not know what has been determined. Instead, one has to adopt the effective theory that one has free will and that one is responsible for one's actions. This theory is not very good at predicting human behavior, but we adopt it because there is no chance of solving the equations arising from the fundamental laws. There is also a Darwinian reason that we believe in free will: A society in which the individual feels responsible for his or her actions is more likely to work together and survive to spread its values. Of course, ants work well together. But such a society is static. It cannot respond to unfamiliar challenges or develop new opportunities. A collection of free individuals who share certain mutual aims, however, can collaborate on their common objectives and yet have the flexibility to make innovations. Thus, such a society is more likely to prosper and to spread its system of values.

The concept of free will belongs to a different arena from that of fundamental laws of science. If one tries to deduce human behavior from the laws of science, one gets caught in the logical paradox of self-referencing systems. If what one does could be predicted from the fundamental laws, then the fact of making that prediction could change what happens. It is like the problems one would get into if time travel were possible, which I don't think it ever will be. If you could see what is going to happen in the future, you could change it. If you knew which horse was going to win the Grand National, you could make a fortune by betting on it. But that action would change the odds. One only has to see *Back to the Future* to realize what problems could arise.

This paradox about being able to predict one's actions is closely related to the problem I mentioned earlier: Will the ultimate theory determine that we come to the right conclusions about the ultimate theory? In that case, I argued that Darwin's idea of natural selection would lead us to the correct answer. Maybe the correct answer is not the right way to describe it, but natural selection should at least lead us to a set of physical laws that work fairly well. However, we cannot apply those physical laws to deduce human be-

135

havior for two reasons. First, we cannot solve the equations. Second, even if we could, the fact of making a prediction would disturb the system. Instead, natural selection seems to lead to us adopting the effective theory of free will. If one accepts that a person's actions are freely chosen, one cannot then argue that in some cases they are determined by outside forces. The concept of "almost free will" doesn't make sense. But people tend to confuse the fact that one may be able to guess what an individual is likely to choose with the notion that the choice is not free. I would guess that most of you will have a meal this evening, but you are quite free to choose to go to bed hungry. One example of such confusion is the doctrine of diminished responsibility: the idea that persons should not be punished for their actions because they were under stress. It may be that someone is more likely to commit an antisocial act when under stress. But that does not mean that we should make it even more likely that he or she commit the act by reducing the punishment.

One has to keep the investigation of the fundamental laws of science and the study of human behavior in separate compartments. One cannot use the fundamental laws to deduce human behavior, for the reasons I have explained. But one might hope that we could employ both the intelligence and the powers of logical thought that we have developed through natural selection. Unfortunately, natural selection has also developed other characteristics, such as aggression. Aggression would have given a survival advantage in cave dweller days and earlier and so would have been favored by natural selection. The tremendous increase in our powers of destruction brought about by modern science and technology, however, has made aggression a very dangerous quality, one that threatens the survival of the whole human race. The trouble is, our aggressive instincts seem to be encoded in our DNA. DNA changes by biological evolution only on a time scale of millions of years, but our powers of destruction are increasing on a time scale for the evolution of information, which is now only twenty or thirty years. Unless we can use our intelligence to control our aggression, there is not much chance for the human race. Still, while there's life, there's hope. If we can survive the next hundred years or so, we will have spread to other planets and possibly to other stars. This will make it much less likely that the entire human race will be wiped out by a calamity such as a nuclear war.

To recapitulate: I have discussed some of the problems that arise if one believes that everything in the universe is determined. It doesn't make much difference whether this determinism is due to an omnipotent God or to the laws of science. Indeed, one could always say that the laws of science are the expression of the will of God.

I considered three questions: First, how can the complexity of the universe and all its trivial details be determined by a simple set of equations? Alternatively, can one really believe that God chose all the trivial details, like who should be on the cover of *Cosmopolitan*? The answer seems to be that the uncertainty principle of quantum mechanics means that there is not just a single history for the universe but a whole family of possible histories. These histories may be similar on very large scales, but they will differ greatly on normal, everyday scales. We happen to live on one particular history that has certain properties and details. But there are very similar intelligent beings who live on histories that differ in who won the war and who is Top of the Pops. Thus, the trivial details of our universe arise because the fundamental laws incorporate quantum mechanics with its element of uncertainty or randomness.

The second question was: If everything is determined by some fundamental theory, then what we say about the theory is also determined by the theory—and why should it be determined to be correct, rather than just plain wrong or irrelevant? My answer to this was to appeal to Darwin's theory of natural selection: Only those individuals who drew the appropriate conclusions about the world around them would be likely to survive and reproduce.

The third question was: If everything is determined, what becomes of free will and our responsibility for our actions? But the only objective test of whether an organism has free will is whether its behavior can be predicted. In the case of human beings, we are quite unable to use the fundamental laws to predict what people will do, for two reasons. First, we cannot solve the equations for the very large number of particles involved. Second, even if we could solve the equations, the fact of making a prediction would disturb the system and could lead to a different outcome. So as we cannot predict human behavior, we may as well adopt the effective theory that humans are free agents who can choose what to do. It seems that there are definite survival advantages to believing in free will and responsibility for one's actions. That means this belief should be reinforced by natural selection. Whether the language-transmitted sense of responsibility is sufficient to control the DNA-transmitted instinct of aggression remains to be seen. If it does not, the human race will have been one of natural selection's dead ends. Maybe some other race of intelligent beings elsewhere in the galaxy will achieve a better balance between responsibility and aggression. But if so, we might have expected to be contacted by them, or at least to detect their radio signals. Maybe they are aware of our existence but don't want to reveal themselves to us. That might be wise, given our record.

In summary, the title of this essay was a question: Is everything determined? The answer is yes, it is. But it might as well not be, because we can never know what is determined.

Thirteen

THE FUTURE
OF THE
UNIVERSE*

T HE SUBJECT OF this essay is the future of the universe,
or rather, what scientists think the future will be. Of
course, predicting the future is very difficult. I once thought I
should write a book called *Yesterday's Tomorrow: A History of
the Future*. It would have been a history of predictions of the
future, nearly all of which have fallen very wide of the mark.
But despite these failures, scientists still think that they can predict the future.

In earlier times foretelling the future was the job of oracles
or sibyls. These were often women, who would be put into a

*Darwin lecture given at the University of Cambridge in January 1991.

trance by some drug or by breathing the fumes from a volcanic vent. Their ravings would then be interpreted by the surrounding priests. The real skill lay in the interpretation. The famous oracle at Delphi, in ancient Greece, was notorious for hedging its bets or being ambiguous. When the Spartans asked what would happen when the Persians attacked Greece, the oracle replied: Either Sparta will be destroyed, or its king will be killed. I suppose the priests reckoned that if neither of these eventualities happened, the Spartans would be so grateful to Apollo that they would overlook the fact that his oracle had been wrong. In fact, the king was killed defending the pass at Thermopylae in an action that saved Sparta and led to the ultimate defeat of the Persians.

On another occasion, Croesus, King of Lydia, the richest man in the world, asked what would happen if he invaded Persia. The answer was: A great kingdom will fall. Croesus thought this meant the Persian Empire, but it was his own kingdom that fell, and he himself ended up on a pyre, about to be burned alive.

Recent prophets of doom have been more ready to stick their necks out by setting definite dates for the end of the world. These have even tended to depress the stock market, though it beats me why the end of the world should make one want to sell shares for money. Presumably, you can't take either with you.

Thus far, all of the dates set for the end of the world have passed without incident. But the prophets have often had an explanation for their apparent failures. For example, William Miller, the founder of the Seventh-Day Adventists, predicted that the Second Coming would occur between March 21, 1843, and March 21, 1844. When nothing happened, the date was revised to October 22, 1844. When that passed without incident, a new interpretation was put forward. According to this, 1844 was the start of the Second Coming—but first, the names in the Book of Life had to be counted. Only then would the Day of Judgment come for those not in the Book. Fortunately, the counting seems to be taking a long time.

Of course, scientific predictions may not be any more reliable than those of oracles or prophets. One has only to think of weather forecasts. But there are certain situations in which we think that we can make reliable predictions, and the future of the universe, on a very large scale, is one of them.

Over the last three hundred years, we have discovered the scientific laws that govern matter in all normal situations. We still don't know the exact laws that govern matter under very extreme conditions. Those laws are important for understanding how the universe began, but they do not affect the future evolution of the universe, unless and until the universe recollapses to a high-density state. In fact, it is a measure of how little these high-energy laws affect the universe now that we have to spend large amounts of money to build giant particle accelerators to test them.

Even though we may know the relevant laws that govern the universe, we may not be able to use them to predict far into the future. This is because the solutions to the equations of physics may exhibit a property known as chaos. What this means is that the equations may be unstable: Make a slight change to the way a system is by a small amount at one time, and the later behavior of the system may soon become completely different. For example, if you slightly change the way you spin a roulette wheel, you will change the number that comes up. It is practically impossible to predict the number that will come up; otherwise, physicists would be making a fortune at the casinos.

With unstable and chaotic systems, there is generally a time scale on which a small change in an initial state will grow into a change that is twice as big. In the case of the earth's atmosphere, this time scale is of the order of five days, about the time it takes for air to blow right around the world. One can make reasonably accurate weather forecasts for periods up to five days, but to predict the weather much further ahead would require both a very accurate knowledge of the present state of the atmosphere *and* an impossibly complicated calculation. There is no way that we can predict the weather six months ahead, beyond giving the seasonal average.

We also know the basic laws that govern chemistry and biology, so in principle we ought to be able to determine how the brain works. But the equations that govern the brain almost certainly have chaotic behavior, in that a very small change in the initial state can lead to a very different outcome. Thus, in practice we cannot predict human behavior, even though we know the equations that govern it. Science cannot predict the future of human society or even if it has any future. The danger is that our power to damage or destroy the environment or one another is increasing much more rapidly than our wisdom in using this power.

Whatever happens on earth, the rest of the universe will carry on regardless. It seems that the motion of the planets around the sun is ultimately chaotic, though with a long time scale. This means that the errors in any prediction get bigger as time goes on. After a certain time, it becomes impossible to predict the motion in detail. We can be fairly sure that the earth will not have a close encounter with Venus for quite a long time, but we cannot be certain that small perturbations in the orbits could not add up to cause such an encounter a billion years from now. The motion of the sun and other stars around the galaxy, and of the galaxy in the local group of galaxies, is also chaotic.

We observe that other galaxies are moving away from us, and the farther they are from us, the faster they are moving away. This means that the universe is expanding in our neighborhood: The distances between different galaxies are increasing with time.

Evidence that this expansion is smooth and not chaotic is given by a background of microwave radiation that we observe coming from outer space. You can actually observe this radiation yourself by tuning your television to an empty channel. A small percent of the flecks you see on the screen are due to microwaves from beyond the solar system. It is the same kind of radiation that you get in a microwave oven, but very much weaker. It would only raise food to 2.7 degrees above absolute zero, so it is not much good for warming up your take-away pizza. This radiation is thought to be left over from a hot early stage of the universe. But the most remarkable thing about it is that the amount of radiation seems to be very nearly the same from every direction. This radiation has been measured very accurately by the Cosmic Background Explorer satellite. A map of the sky made from these observations would show different temperatures of radiation. These temperatures are different in different directions, but the variations are very small, only one part in a hundred thousand. There have to be some differences in the microwaves from different directions because the universe is not completely smooth; there are local irregularities like stars, galaxies, and clusters of galaxies. But the variations in the microwave background are as small as they possibly can be, compatible with the local irregularities that we observe. To 99,999 parts out of 100,000, the microwave background is the same in every direction.

In ancient times, people believed that the earth was at the center of the universe. They would therefore not have been surprised that the background was the same in every direction.

Since the time of Copernicus, however, we have been demoted to a minor planet going around a very average star in the outer edge of a typical galaxy that is only one of a hundred billion galaxies we can see. We are now so modest that we cannot claim any special position in the universe. We must therefore assume that the background is also the same in any direction about any other galaxy. This is possible only if the average density of the universe and the rate of expansion are the same everywhere. Any variation in the average density, or the rate of expansion, over a large region would cause the microwave background to be different in different directions. This means that on a very large scale, the behavior of the universe is simple and is not chaotic. It can therefore be predicted far into the future.

Because the expansion of the universe is so uniform, one can describe it in terms of a single number, the distance between two galaxies. This is increasing at the present time, but one would expect the gravitational attraction between different galaxies to be slowing down the rate of expansion. If the density of the universe is greater than a certain critical value, gravitational attraction will eventually stop the expansion and make the universe start to contract again. The universe would collapse to a big crunch. This would be rather like the big bang that began the universe. The big crunch would be what is called a singularity, a state of infinite density at which the laws of physics would break down. This means that even if there were events after the big crunch, what happened at them could not be predicted. But without a causal connection between events, there is no meaningful way that one can say that one event happened after another. One might as well say that our universe came to an end at the big crunch and that any events that occurred "after" were part of another, separate universe. It is a bit like reincarnation. What meaning can one give to the claim that a new baby is the same as someone who died if the baby doesn't inherit any characteristics or memories from its previous life? One might as well say that it is a different individual.

If the average density of the universe is less than the critical value, it will not recollapse but will continue to expand forever. After a certain time the density will become so low that gravitational attraction will not have any significant effect on slowing down the expansion. The galaxies will continue to move apart at a constant speed.

So the crucial question for the future of the universe is: What is the average density? If it is less than the critical value, the universe will expand forever. But if it is greater, the universe will recollapse and time itself will come to an end at the big crunch. I do, however, have certain advantages over other prophets of doom. Even if the universe is going to recollapse, I can confidently predict that it will not stop expanding for at least ten billion years. I don't expect to be around to be proved wrong.

We can try to estimate the average density of the universe from observations. If we count the stars that we can see and add up their masses, we get less than one percent of the critical density. Even if we add in the masses of the clouds of gas that we observe in the universe, it still brings the total up to only about one percent of the critical value. However, we know that the universe must also contain what is called dark matter, which we cannot observe directly. One piece of evidence for this dark matter comes from spiral galaxies. These are enormous pancake-shaped collections of stars and gas. We observe that they are rotating about their centers, but the rate of rotation is sufficiently high that they would fly apart if they contained only the stars and gas that we observe. There must be some unseen form of matter whose gravitational attraction is great enough to hold the galaxies together as they rotate.

Another piece of evidence for dark matter comes from clusters of galaxies. We observe that galaxies are not uniformly distributed throughout space; they are gathered together in clusters that range from a few galaxies to millions. Presumably, these clusters are formed because the galaxies attract each other into groups. However, we can measure the speeds at which individual galaxies are moving in these clusters. We find they are so high that the clusters would fly apart unless they were held together by gravitational attraction. The mass required is considerably greater than the masses of all the galaxies. This is the case even if we take the galaxies to have the masses required to hold themselves together as they rotate. It follows, therefore, that there must be extra dark matter present in clusters of galaxies outside the galaxies that we see.

One can make a fairly reliable estimate of the amount of dark matter in those galaxies and clusters for which we have definite evidence. But this estimate is still only about ten percent of the critical density needed to cause the universe to collapse again. Thus, if one just went by the observational evidence, one would predict that the universe would continue to expand forever. After another five billion years or so, the sun would reach the end of its nuclear fuel. It would swell up into what is called a red giant until it swallowed up the earth and the other nearer planets. It would then settle down to be a white dwarf star a few thousand miles across. So I am predicting the end of the world, but not just yet. I don't think this prediction will depress the stock market too much. There are one or two more immediate problems on the horizon. In any event, by the time the sun blows up, we should have mastered the art of interstellar travel, provided we have not already destroyed ourselves.

After ten billion years or so, most of the stars in the universe will have burned out. Stars with masses like that of the sun will become either white dwarfs or neutron stars, which are even smaller and denser than white dwarfs. More massive stars can become black holes, which are still smaller and have a strong gravitational field that no light can escape. However, these remnants will still continue to go around the center of our galaxy about once every hundred million years. Close encounters between the remnants will cause a few to be flung right out of the galaxy. The remainder will settle down to closer orbits about the center and will eventually collect together to form a giant black hole at the center of the galaxy. Whatever the dark matter in galaxies and clusters is, it might also be expected to fall into these very large black holes.

It could be assumed, therefore, that most of the matter in galaxies and clusters would eventually end up in black holes. However, some time ago I discovered that black holes aren't as black as they have been painted. The uncertainty principle of quantum mechanics says that particles cannot have both a well-defined position and a well-defined speed. The more accurately the position of the particle is defined, the less accurately its speed can be defined, and vice versa. If a particle is in a black hole, its position is well-defined to be within the black hole. This means that its speed cannot be exactly defined. It is therefore possible for the speed of the particle to be greater than the speed of light. This would enable it to escape from the black hole. Particles and radiation will thus slowly leak out of a black hole. A giant black hole at the center of a galaxy would be millions of miles across. Thus, there would be a large uncertainty in the position of a particle inside it. The uncertainty in the particle's speed would therefore be small, which means that it would take a very long time for a particle to escape from the black hole. But it would eventually. A large black hole at the center of a galaxy could take 10^{90} years to evaporate away and disappear completely; that is, a one followed by ninety zeroes. This is far longer than the present age of the universe, which is a mere 10^{10} years; a one followed by ten zeroes. Still, there will be plenty of time, if the universe is going to expand forever.

The future of a universe that expanded forever would be rather boring. But it is by no means certain that the universe will expand forever. We have definite evidence only for about one-tenth of the density needed to cause the universe to re-collapse. Still, there might be further kinds of dark matter that we have not detected that could raise the average density of the universe to the critical value or above it. This additional dark matter would have to be located outside galaxies and clusters of galaxies. Otherwise, we would have noticed its effect on the rotation of galaxies or the motions of galaxies in clusters.

Why should we think there might be enough dark matter to make the universe recollapse eventually? Why don't we just believe in the matter for which we have definite evidence? The reason is that having even a tenth of the critical density now requires an incredibly careful choice of the initial density and rate of expansion. If the density of the universe one second after the big bang had been greater by one part in a thousand billion, the universe would have recollapsed after ten years. On the other hand, if the density of the universe at that time had been less by the same amount, the universe would have been essentially empty since it was about ten years old.

How is it that the initial density of the universe was chosen so carefully? Maybe there is some reason that the universe should have precisely the critical density. There seem to be two possible explanations. One is the so-called anthropic principle, which can be paraphrased as: The universe is as it is because if it were different, we wouldn't be here to observe it. The idea is that there could be many different universes with different densities. Only those that are very close to the critical density would last for long enough and contain enough matter for stars and planets to form. Only in those universes will there be intelligent beings to ask the question: Why is the density so close to the critical density? If this is the explanation of the present density of the universe, there is no reason to believe that the universe contains more matter than we have already detected. A tenth of the critical density would be enough matter for galaxies and stars to form.

Many people do not like the anthropic principle, however, because it seems to attach too much importance to our own existence. There thus has been a search for another possible explanation of why the density should be so close to the critical value. This search has led to the theory of inflation in the early universe. The idea is that the size of the universe may have kept doubling, in the same way that prices double every few months in countries undergoing extreme inflation. However, the inflation of the universe would have been much more rapid and more extreme: an increase by a factor of at least a billion billion billion, in a tiny inflation, would have caused the universe to have so nearly the exact critical density that it would still be very near the critical density now. Thus, if the theory of inflation is correct, the universe must contain enough dark matter to bring the density up to the critical density. This means that the universe would probably recollapse eventually but not for much longer than the fifteen billion years or so that it has already been expanding.

What could the extra dark matter be that must be there if the theory of inflation is correct? It seems that it is probably different from normal matter, the kind that makes up stars and planets. We can calculate the amounts of various light elements that would have been produced in the hot early stages of the universe in the first three minutes after the big bang. The amounts of these light elements depend on the amount of normal matter in the universe. One can draw a graph showing the amount of light elements vertically and the amount of normal matter in the universe along the horizontal axis. One gets good agreement with the observed abundances if the total amount of normal matter is only about one-tenth of the critical amount now. It could be that these calculations are wrong, but the fact that we get the observed abundances for several different elements is quite impressive.

If there is a critical density of dark matter, the main candidates for what it might be would be remnants left over from the early stages of the universe. One possibility is elementary particles. There are several hypothetical candidates, particles we think might exist but that we have not actually detected yet. But the most promising case is a particle for which we have good evidence, the neutrino. This was thought to have no mass of its own, but some recent observations have suggested that the neutrino may have a small mass. If this is confirmed and found to be of the right value, neutrinos would provide enough mass to bring the density of the universe up to the critical value.

Another possibility is black holes. It is possible that the early universe underwent what is called a phase transition. The boiling and freezing of water are examples of phase transitions. In a phase transition an initially uniform medium, like water, develops irregularities, which in the case of water can be lumps of ice or bubbles of steam. These irregularities might collapse to form black holes. If the black holes were very small, they would have evaporated by now because of the effects of the quantum mechanical uncertainty principle, as described earlier.

But if they were over a few billion tons (the mass of a mountain), they would still be around today and would be very difficult to detect.

The only way we could detect dark matter that was uniformly distributed throughout the universe would be by its effect on the expansion of the universe. One can determine how fast the expansion is slowing down by measuring the speed at which distant galaxies are moving away from us. The point is that we are observing these galaxies in the distant past, when light left them on its journey to us. One can plot a graph of the speed of the galaxies against their apparent brightness or magnitude, which is a measure of their distance from us. Different lines on this graph correspond to different rates of slowing of the expansion. A graph that bends up corresponds to a universe that will recollapse. At first sight the observations seem to indicate recollapse. But the trouble is, the apparent brightness of a galaxy is not a very good indication of its distance from us. Not only is there considerable variation in the intrinsic brightness of galaxies, but there is also evidence that their brightness is varying with time. Since we do not know how much to allow for the evolution of brightness, we can't yet say what the rate of slowing down is: whether it is fast enough for the universe to recollapse eventually, or whether it will continue to expand forever. That will have to wait until we develop better ways of measuring the distances of galaxies. But we can be sure that the rate of slowing down is not so rapid that the universe will collapse in the next few billion years.

Neither expanding forever nor recollapsing in a hundred billion years or so is a very exciting prospect. Isn't there something we can do to make the future more interesting? One way that would certainly do that would be to steer ourselves into a black hole. It would have to be a fairly big black hole, more than a million times the mass of the sun. But there is a good chance there's a black hole that big at the center of our galaxy.

We are not quite sure what happens inside a black hole. There are solutions of the equations of general relativity that would allow one to fall into a black hole and come out of a white hole somewhere else. A white hole is the time reverse of a black hole. It is an object that things can come out of but nothing can fall into. The white hole could be in another part of the universe. This would seem to offer the possibility of rapid intergalactic travel. The trouble is it might be too rapid. If travel through black holes were possible, there would seem nothing to prevent you from arriving back before you set off. You could then do something, like kill your mother, that would have prevented you from going in the first place.

Perhaps fortunately for our survival (and that of our mothers), it seems that the laws of physics do not allow such time travel. There seems to be a Chronology Protection Agency that makes the world safe for historians by preventing travel into the past. What seems to happen is that the effects of the uncertainty principle would cause there to be a large amount of radiation if one traveled into the past. This radiation would either warp space-time so much that it would not be possible to go back in time, or it would cause space-time to come to an end in a singularity like the big bang and the big crunch. Either way, our past would be safe from evil-minded persons. The Chronology Protection Hypothesis is supported by some recent calculations that I and other people have done. But the best evidence we have that time travel is not possible, and never will be, is that we have not been invaded by hordes of tourists from the future.

To sum up: Scientists believe that the universe is governed by well-defined laws that in principle allow one to predict the future. But the motion given by the laws is often chaotic. This means that a tiny change in the initial situation can lead to change in the subsequent behavior that rapidly grows large. Thus, in practice, one can often predict accurately only a fairly short time into the future. However, the behavior of the universe on a very large scale seems to be simple, and not chaotic. One can therefore predict whether the universe will expand forever or whether it will recollapse eventually. This depends on the present density of the universe. In fact, the present density seems to be very close to the critical density that separates recollapse from indefinite expansion. If the theory of inflation is correct, the universe is actually on the knife edge. So I am in the well-established tradition of oracles and prophets of hedging my bets by predicting both ways.

Fourteen

DESERT ISLAND DISCS:

AN INTERVIEW

T*HE BBC's* Desert Island Discs *began broadcasting in 1942 and is the longest-running record program on radio; by now, it is something of a national institution in Britain. Over the years the range of its guests has been enormous. The program has interviewed writers, actors, musicians, film actors and directors, sports figures, comedians, chefs, gardeners, teachers, dancers, politicians, royalty, cartoonists—and scientists. The guests, always referred to as castaways, are asked to choose which eight records they would take with them if they were marooned alone on a desert island. They are also asked to name a luxury object (which must be inanimate) and a book*

157

to accompany them (it is assumed that an appropriate religious text—the Bible, the Koran, or an equivalent volume—is already there, together with the works of Shakespeare). It is taken for granted that the means to play the records exists; the early announcements introducing the program used to say " . . . assuming there is a gramophone and an inexhaustible supply of needles to play them." Today a solar-powered CD player is presumed to be the available means of hearing them.

The program is broadcast weekly, and the guests' choice of records is played during the interview, which normally runs for forty minutes. However, this interview with Stephen Hawking, which was broadcast on Christmas Day 1992, was an exception and ran longer than that.

The interviewer is Sue Lawley.

SUE: In many ways, of course, Stephen, you are already familiar with the isolation of a desert island, cut off from normal physical life and deprived of any natural means of communication. How lonely is it for you?

STEPHEN: I don't regard myself as cut off from normal life, and I don't think people around me would say I was. I don't feel a disabled person—just someone with certain malfunctions of my motor neurones, rather as if I were color blind. I suppose my life can hardly be described as usual, but I feel it is normal in spirit.

SUE: Nevertheless, you have already proved to yourself, unlike most castaways on *Desert Island Discs,* that you are mentally and intellectually self-sufficient, that you've got enough theories and inspiration to keep yourself occupied.

STEPHEN: I suppose I'm naturally a bit introverted, and my difficulties in communication have forced me to rely on myself. But I was a great talker as a boy. I need discussion with other people to stimulate me. I find it a great help in my work to describe my ideas to others. Even if they don't offer any suggestions, the mere fact of having to organize my thoughts so that I can explain them to others often shows me a new way forward.

SUE: But what about emotional fulfillment, Stephen? Even a brilliant physicist must need other people to find that.

STEPHEN: Physics is all very well, but it is completely cold. I couldn't carry on with my life if I only had physics. Like everyone else, I need warmth, love, and affection. Again, I'm very fortunate, much more fortunate than many people with my disabilities, in receiving a great deal of love and affection. Music is also very important to me.

SUE: Tell me, which gives you greater pleasure, physics or music?

STEPHEN: I have to say that the pleasure I have had when everything works out in physics is more intense than I have ever had with music. But things work out like that only a few times in one's career, whereas one can put on a disc when-ever one wants.

SUE: And the first record you'd play on your desert island?

STEPHEN: *Gloria,* by Poulenc. I first heard it last summer in Aspen, Colorado. Aspen is primarily a ski resort, but in the summer they have physics meetings. Next door to the physics center is an enormous tent where they hold a music festival. As you sit working out what happens when black holes evaporate, you can hear the rehearsals. It is ideal; it combines my two main pleasures, physics and music. If I can have both on my desert island, I won't want to be rescued. Not, that is, until I have made a discovery in theoretical physics that I want to tell everyone about. I suppose a satellite dish, so I could get physics papers by electronic mail, would be against the rules.

SUE: Radio can hide physical shortcomings, but on this occasion it's disguising something else. Back seven years ago, Stephen, you literally lost your voice. Can you tell me what happened?

STEPHEN: I was in Geneva, at CERN, the big particle accelerator, in the summer of 1985. I was intending to go on to Bayreuth, in Germany, to hear Wagner's *Ring* cycle of operas. But I caught pneumonia and was rushed to hospital. The hospital in Geneva suggested to my wife that it was not worth keeping the life support machine on. But she was not having any of that. I was flown back to Addenbrookes Hospital in Cambridge, where a surgeon called Roger Grey carried out a tracheostomy. That operation saved my life but took away my voice.

SUE: But your speech was in any case by then very slurred and difficult to understand, wasn't it? So presumably the power of speech would have deserted you eventually anyway, wouldn't it?

STEPHEN: Although my voice was slurred and difficult to comprehend, the people close to me could still understand me. I could give seminars through an interpreter, and I could dictate scientific papers. But for a time after my operation, I was devastated. I felt that if I couldn't get my voice back, it wasn't worth carrying on.

SUE: Then a California computer expert read about your plight and sent you a voice. How does it work?

STEPHEN: His name was Walt Woltosz. His mother-in-law had had the same condition as me, so he had developed a computer program to help her communicate. A cursor moves across a screen. When it is on the option you want, you operate a switch by head or eye movement, or in my case by hand. In this way, one can select words that are printed out on the lower half of the screen. When one has built up what one wants to say, one can send it to a speech synthesizer or save it on disk.

SUE: But it's a slow business.

STEPHEN: It is slow, roughly one-tenth the speed of normal speech. But the speech synthesizer is so much clearer than I was previously. British people describe its accent as American, but the Americans say it is Scandinavian or Irish. Anyway, whatever it is, everyone can understand it. My elder children adapted to my natural voice as it got worse, but my youngest son, who was only six at the time of my tracheostomy, never could make me out before. Now he has no difficulty. That means a great deal to me.

SUE: It also means you can demand good notice of any interviewer's questions and need only answer when you're good and ready, doesn't it?

STEPHEN: For long, recorded programs like this, it helps to have advance notice of the questions, so I don't use hours and hours of recording tape. In a way that gives me more control. But I really prefer to answer questions off the cuff. I do that after seminars and popular lectures.

SUE: But as you say, the process means that you have control, and I know that's quite important to you. Your family and friends sometimes call you stubborn or bossy. Do you plead guilty to being those things?

STEPHEN: Anyone with any nous is called stubborn at times. I would prefer to say I'm determined. If I hadn't been fairly determined, I wouldn't be here now.

SUE: Were you always like that?

STEPHEN: I just want to have the same degree of control over my life that other people have. Far too often, disabled people have their lives ruled by others. No able-bodied person would put up with it.

SUE: Let's have your second record.

STEPHEN: The Brahms Violin Concerto. This was the first LP I bought. It was 1957, and 33 rpm records had recently appeared in Britain. My father would have regarded it as recklessly self-indulgent to buy a record player, but I persuaded him I could assemble one from parts that I could buy cheap. That appealed to him as a Yorkshireman. I housed the turntable and amplifier in the case of an old 78 gramophone. If I had kept it, it would now be very valuable.

Having built this record player, I needed something to play on it. A school friend suggested the Brahms Violin Concerto, as no one in our circle at school had a record of it. I remember it cost thirty-five shillings, which was a lot in those days, especially to me. Record prices have gone up, but they are now a lot less in real terms.

When I first heard this record in a shop, I thought it sounded rather strange and I wasn't sure I liked it, but I felt I had to say I did. However, over the years it has come to mean a great deal to me. I would like to play the start of the slow movement.

SUE: An old family friend has said that your family, when you were a boy, was, and I quote, "highly intelligent, very clever, and very eccentric." Looking back, do you think that's a fair description?

STEPHEN: I can't comment on whether my family were intelligent, but we certainly didn't feel we were eccentric. However, I suppose we may have seemed so by the standards of St. Albans, which was a pretty staid place when we lived there.

SUE: And your father was a specialist in tropical diseases.

STEPHEN: My father did research in tropical medicine. He quite often went to Africa, to try out new drugs in the field.

SUE: So was your mother the greater influence on you, and if so, how would you characterize that influence?

STEPHEN: No, I would say my father was the greater influence. I modeled myself on him. Because he was a scientific researcher, I felt that scientific research was the natural thing to do when one grew up. The only difference was that I was not attracted to medicine or biology because they seemed too inexact and descriptive. I wanted something more fundamental, and I found it in physics.

SUE: Your mother has said that you always had what she described as a strong sense of wonder. "I could see that the stars could draw him," she said. Do you remember that?

STEPHEN: I remember coming home late one night from London. In those days they turned the streetlights out at midnight, to save money. I saw the night sky as I had never seen it before, with the Milky Way going right across. There won't be any streetlights on my desert island, so I should get a good view of the stars.

SUE: Obviously you were very bright as a child, you were very competitive in games at home with your sister, but you could come practically at the bottom of the class at school and not care about it at all, couldn't you?

STEPHEN: That was in my first year at St. Albans school. But I should say that it was a very bright class, and I did much better in examinations than in classwork. I was sure that I really could do well—it was just my handwriting and general untidiness that caused me to be placed so low.

SUE: Record number three?

STEPHEN: When I was an undergraduate at Oxford, I read Aldous Huxley's novel *Point Counterpoint*. This was intended as a portrait of the 1930s and had an enormous cast of characters. Most of these were pretty cardboard, but there was one who was rather more human and was obviously modeled on Huxley himself. This man killed the leader of the British Fascists, a character based on Sir Oswald Mosley. He then let the Party know he had done it and put on the gramophone records of Beethoven's String Quartet, Opus 132. In the middle of the third movement he answered the door and was shot by the Fascists.

It really is a very bad novel, but Huxley was right about his choice of music. If I knew that a tidal wave was on the way to overwhelm my desert island, I would play the third movement of this quartet.

SUE: You went up to Oxford, to University College, to read maths and physics, where you worked by your own calculations an average of about an hour a day. Although it has to be said you rowed, drank beer, and played silly tricks on people with some pleasure, according to what I've read. What was the problem? Why couldn't you be bothered to work?

STEPHEN: It was the end of the fifties, and most young people were disillusioned with what was called the Establishment. There seemed nothing to look forward to but affluence and more affluence. The Conservatives had just won their third election victory with the slogan, "You've never had it so good." I and most of my contemporaries were bored with life.

SUE: Nevertheless, you still managed to solve in a few hours problems that your fellow students couldn't do in as many weeks. *They* were obviously aware, from what they've said since, that you had an exceptional talent. Were you aware, do you think?

STEPHEN: The physics course at Oxford at that time was ridiculously easy. One could get through without going to any lectures, just by going to one or two tutorials a week. You didn't need to remember many facts, just a few equations.

SUE: But it was at Oxford, wasn't it, that you first noticed that your hands and feet weren't quite doing what you wanted them to do. How did you explain that to yourself at that time?

STEPHEN: In fact, the first thing I noticed was that I couldn't row a sculling boat properly. Then I had a bad fall down the stairs from the junior common room. I went to the college doctor after the fall because I was worried that I might have brain damage, but he thought there was nothing wrong and told me to cut down on the beer. After my finals at Oxford, I went to Persia for the summer. I was definitely weaker when I came back, but I thought that was caused by a bad stomach upset that I had had.

SUE: But at what point did you give in and admit that there was something really wrong and decide to get medical advice?

STEPHEN: I was at Cambridge by then, and I went home at Christmas. That was the very cold winter of '62 to '63. My mother persuaded me to go and skate on the lake in St. Albans, even though I knew I was not really up to it. I fell over and had great difficulty getting up. My mother realized there was something wrong. She took me to the family doctor.

SUE: And then three weeks in hospital, and they told you the worst?

STEPHEN: In fact, it was Barts Hospital in London, because my father was a Barts man. I was in for two weeks, having tests, but they never actually told me what was wrong, except that it was not MS and that it was not a typical case. They didn't tell me what the prospects were, but I guessed enough to know they were pretty bad, so I didn't want to ask.

SUE: And finally, in fact, you were told that you had only a couple of years or so to live. Let's pause at that point in your story, Stephen, and have your next record.

STEPHEN: *The Valkyrie,* act one. This was another early LP, with Melchior and Lehmann. It was originally recorded on 78s before the war and transferred to an LP in the early sixties. After I was diagnosed with motor neurone disease in 1963, I turned to Wagner as someone who suited the dark and apocalyptic mood I was in. Unfortunately, my speech synthesizer is not very well-educated, and it pronounces him with a soft *W.* I have to spell him V-A-R-G-N-E-R to get it to sound approximately right.

The four operas of the *Ring* cycle are Wagner's greatest work. I went to see them at Bayreuth, in Germany, with my sister Philippa in 1964. I didn't know the *Ring* well at that time, and *The Valkyrie,* the second opera in the cycle, made a tremendous impression on me. It was a production by Wolfgang Wagner, and the stage was almost totally dark. It is the love story of twins, Siegmund and Sieglinde, who were separated in childhood. They meet again when Siegmund takes refuge in the house of Hunding, Sieglinde's husband and Siegmund's enemy. The excerpt I have chosen is Sieglinde's account of her forced wedding to Hunding. In the middle of the celebrations, an old man comes into the hall. The orchestra plays the Valhalla motif, one of the noblest themes in the *Ring,* because he is Wotan, the leader of the gods and the father of Siegmund and Sieglinde. He plunges a sword into the trunk of a tree. The sword is intended for Siegmund. At the end of the act Siegmund draws it out, and the two run off into the forest.

SUE: Reading about you, Stephen, it almost seems as if that death sentence, being told you had only a couple of years or so to live, woke you up, if you like, made you concentrate on life.

STEPHEN: Its first effect was to depress me. I seemed to be getting worse fairly rapidly. There didn't seem any point in doing anything or working on my Ph.D. because I didn't know I would live long enough to finish it. But then things started to improve. The condition developed more slowly, and I began to make progress in my work, particularly in showing that the universe must have had a beginning in a big bang.

SUE: You've even said in one interview that you thought you are happier now than before you got ill.

STEPHEN: I certainly am happier now. Before I got motor neurone disease, I was bored with life. But the prospect of an early death made me realize life is really worth living. There is so much one can do, so much that anyone can do. I have a real feeling of achievement that I have made a modest but significant contribution to human knowledge despite my condition. Of course, I'm very fortunate, but everyone can achieve something if they try hard enough.

SUE: Would you go so far as to say that you mightn't have achieved all you have, had you not had motor neurone disease, or is that just too simplistic?

STEPHEN: No, I don't think motor neurone disease can be an advantage to anyone. But it was less of a disadvantage to me than to other people, because it didn't stop me doing what I wanted, which was to try and understand how the universe operates.

SUE: Your other inspiration, when you were trying to come to terms with the disease, was a young woman called Jane Wilde, whom you'd met at a party and fallen in love with and subsequently married. How much of your success, would you say, do you owe to her, to Jane?

STEPHEN: I certainly wouldn't have managed it without her. Being engaged to her lifted me out of the slough of despond I was in. And if we were to get married, I had to get a job and I had to finish my Ph.D. I began to work hard and found I enjoyed it. Jane looked after me single-handedly as my condition got worse. At that stage, no one was offering to help us, and we certainly couldn't afford to pay for help.

SUE: And together you defied the doctors, not only because you went on living but also because you had children. You had Robert in 1967, Lucy in 1970, and then Timothy in 1979. How shocked were the doctors?

STEPHEN: In fact, the doctor who diagnosed me washed his hands of me. He felt that there was nothing that could be done. I never saw him after the initial diagnosis. In effect, my father became my doctor, and it was to him I turned for advice. He told me there was no evidence the disease was hereditary. Jane managed to look after me and two children. It was only when we went to California in 1974 that we had to get outside help, first a student living with us, and later nurses.

SUE: But now you and Jane aren't together anymore.

STEPHEN: After my tracheostomy operation I needed twenty-four-hour nursing. That put a greater and greater strain on the marriage. Eventually I moved out, and I now live in a new flat in Cambridge. We now live separately.

SUE: Let's have some more music.

STEPHEN: The Beatles, "Please Please Me." After my first four rather serious choices, I would need some light relief. For me and many others, the Beatles came as a welcome breath of fresh air to a rather stale and sickly pop scene. I used to listen to the top twenty on Radio Luxembourg on Sunday evenings.

SUE: Despite all the honors that have been heaped on you, Stephen Hawking—and I should specifically mention that you're Lucasian Professor of Mathematics at Cambridge, Isaac Newton's chair—you decided to write a popular book about your work, for, I think, a very simple reason. You needed the money.

STEPHEN: While I thought I might make a modest amount from a popular book, the main reason I wrote *A Brief History of Time* was because I enjoyed it. I was excited about the discoveries that have been made in the last twenty-five years, and I wanted to tell people about them. I never expected it to do as well as it did.

SUE: Indeed, it's broken all the records and got into the *Guinness Book of Records* for the length of time it's been on the best-seller lists, and it's still there. Nobody seems to know how many copies have been sold worldwide, but it's certainly in excess of ten million. People buy it, obviously, but the question goes on being asked: Do they read it?

STEPHEN: I know Bernard Levin got stuck on page twenty-nine, but I know plenty of people have got further. All over the world, people come up to me and tell me how much they have enjoyed it. They may not have finished it or have understood everything they read. But they have at least got the idea that we live in a universe governed by rational laws that we can discover and understand.

SUE: It was the concept of the black hole that first appealed to the public imagination and attracted renewed interest in cosmology. Did you ever watch all those *Star Trek*s, "to boldly go where no man has ever gone before" and so on, and if so, did you enjoy them?

STEPHEN: I read a lot of science fiction when I was a teenager. But now that I work in the field myself, I find most science fiction a bit facile. It is so easy to write about hyperspace drive or beaming people up, if you don't have to make it part of a consistent picture. Real science is much more exciting because it is actually happening out there. Science fiction writers never suggested black holes before physicists thought of them. But now we have good evidence for a number of black holes.

SUE: What would happen if you fell into a black hole?

STEPHEN: Everyone who reads science fiction knows what happens if you fall into a black hole. You get made into spaghetti. But what is much more interesting is that black holes aren't completely black. They send out particles and radiation at a steady rate. This causes the black hole to evaporate slowly, but what eventually happens to the black hole and its contents is not known. This is an exciting area of research, but science fiction writers have not caught up with it yet.

SUE: And that radiation you mentioned, of course, is called Hawking radiation. It wasn't you who discovered black holes, although you've gone on to prove they're not black. But it was their discovery that made you begin to think more closely about the origins of the universe, wasn't it?

STEPHEN: The collapse of a star to form a black hole is in many ways like the time reverse of the expansion of the universe. A star collapses from a fairly low-density state to one of very high density. And the universe expands from a very high-density state to lower densities. There's an important difference: We are outside the black hole, but we are inside the universe. But both are characterized by thermal radiation.

SUE: You say that it's not known what eventually happens to a black hole and its contents. But I thought the theory was that whatever happened, whatever disappeared into a black hole, including an astronaut, would eventually be recycled as Hawking radiation.

STEPHEN: The mass energy of the astronaut will be recycled as radiation sent out by the black hole. But the astronaut himself, or even the particles of which he is made, won't come out of the black hole. So the question is, what happens to them? Do they get destroyed, or do they pass into another universe? That is something I would dearly like to know, not that I'm thinking of jumping into a black hole.

SUE: Do you work, Stephen, on intuition—that's to say, do you arrive at a theory that you rather like and that appeals to you, and set about proving it? Or as a scientist, do you always have to make your way logically toward a conclusion and not dare attempt to guess it in advance?

STEPHEN: I rely on intuition a great deal. I try to guess a result, but then I have to prove it. And at this stage, I quite often find that what I had thought of is not true or that something else is the case that I had never thought of. That is how I

171

found black holes aren't completely black. I was trying to prove something else.

SUE: More music.

STEPHEN: Mozart has always been one of my favorites. He wrote an incredible amount of music. For my fiftieth birthday earlier this year, I was given his complete works on CD, over two hundred hours of it. I'm still working my way through it. One of the greatest is the *Requiem*. Mozart died before the *Requiem* was finished, and it was completed by one of his students from fragments Mozart had left. The introit we are about to hear is the only part completely written and orchestrated by Mozart.

SUE: To oversimplify your theories hugely, and I hope you'll forgive me for this, Stephen, you once believed, as I understand it, that there was a point of creation, a big bang, but you no longer believe that to be the case. You believe that there was no beginning and there is no end, that the universe is self-contained. Does that mean that there was no act of creation and therefore that there's no place for God?

STEPHEN: Yes, you have oversimplified. I still believe the universe has a beginning in real time, at a big bang. But there's another kind of time, imaginary time, at right angles to real time, in which the universe has no beginning or end. This would mean that the way the universe began would be determined by the laws of physics. One wouldn't have to say that God chose to set the universe going in some arbitrary way that we couldn't understand. It says nothing about whether or not God exists—just that He is not arbitrary.

SUE: But how, if there's a possibility that God doesn't exist, do you account for all those things that are beyond science:

love, and the faith that people have had and have in you, and indeed in your own inspiration?

STEPHEN: Love, faith, and morality belong to a different category to physics. You cannot deduce how one should behave from the laws of physics. But one could hope that the logical thought that physics and mathematics involves would guide one also in one's moral behavior.

SUE: But I think that many people do feel you have effectively dispensed with God. Are you denying that, then?

STEPHEN: All that my work has shown is that you don't have to say that the way the universe began was the personal whim of God. But you still have the question: Why does the universe bother to exist? If you like, you can define God to be the answer to that question.

SUE: Let's have record number seven.

STEPHEN: I'm very fond of opera. I had thought of choosing all my eight discs to be opera, ranging from Gluck and Mozart, through Wagner, to Verdi and Puccini. But in the end I cut it down to two. One had to be Wagner, and eventually I decided the other should be Puccini. *Turandot* is by far his greatest opera, but again, he died before he finished it. The excerpt I have chosen is Turandot's account of how a princess in ancient China was raped and carried away by the Mongols. In revenge for this, Turandot is going to ask her suitors three questions. If they can't answer, they will be executed.

SUE: What does Christmas mean to you?

STEPHEN: It is a bit like the American Thanksgiving, a time to be with one's family and to give thanks for the year past.

It is also the time to look forward to the year ahead, as symbolized by the birth of a child in a stable.

SUE: And to be materialistic about it, what presents have you asked for—or are you so well off these days that you're the man who has everything?

STEPHEN: I prefer surprises. If one asks for something specific, one isn't letting the giver have any freedom or the opportunity to use his or her imagination. But I don't mind it being known that I'm fond of chocolate truffles.

SUE: So far, Stephen, you've lived for thirty years longer than predicted. You've fathered children you were told you'd never have, you've written a best seller, you've turned age-old beliefs about space and time on their heads. What else are you planning to do before you quit this planet?

STEPHEN: All that has been possible only because I've been fortunate enough to receive a great deal of help. I'm pleased with what I have managed to achieve, but there's a great deal more I would like to do before I pass on. I won't talk about my private life, but scientifically, I would like to know how one should unify gravity with quantum mechanics and the other forces of nature. In particular, I want to know what happens to a black hole when it evaporates.

SUE: The last record now.

STEPHEN: I will have to get you to pronounce this. My speech synthesizer is American and is hopeless at French. It is Edith Piaf singing *"Je ne regrette rien."* That just about sums up my life.

SUE: Now, Stephen, if you could take only one of those eight records with you, which one would it be?

STEPHEN: It would have to be the Mozart *Requiem*. I could listen to that until the batteries in my disc Walkman ran out.

SUE: And your book? Of course, the complete works of Shakespeare and the Bible are waiting for you.

STEPHEN: I think I will take *Middlemarch* by George Eliot. I think someone, maybe it was Virginia Woolf, said it was a book for adults. I'm not sure I'm grown up yet, but I will give it a try.

SUE: And your luxury?

STEPHEN: I will ask for a large supply of crème brulée. For me, that is the epitome of luxury.

SUE: Not the chocolate truffles, then: a large supply of crème brulée instead. Dr. Stephen Hawking, thank you very much for letting us hear your Desert Island Discs, and a happy Christmas.

STEPHEN: Thank you for choosing me. I wish you all a happy Christmas from my desert island. I bet I'm having better weather than you.

Index

176

RD6Z